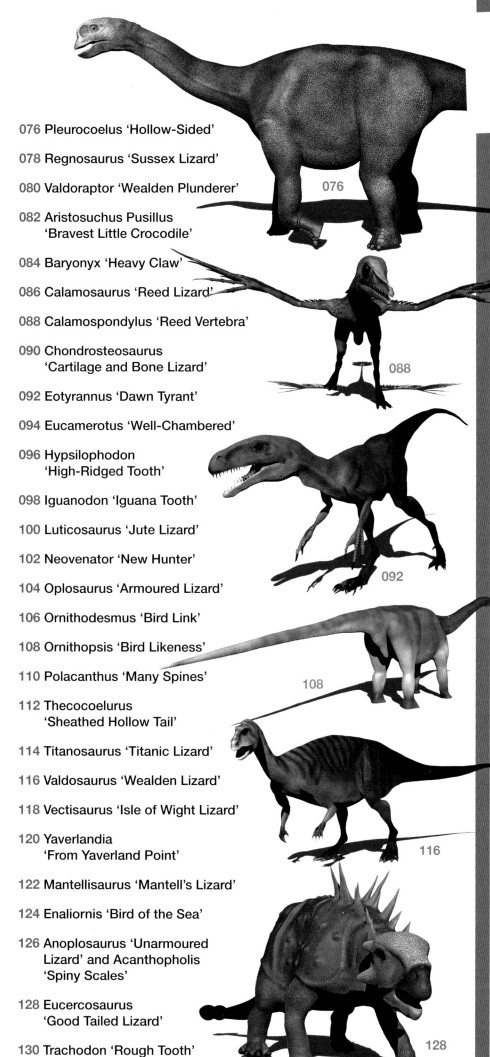

ART: Chris Wardle
(all images: British Geological Survey
© NERC, 2015)

AUTHORS: Tony Carter
and Dan Sharp

COVER DESIGN: Michael Baumber

DESIGN: Panda Media

REPROGRAPHICS:
Jonathan Schofield
and Paul Fincham

SUB EDITOR: Jack Harrison

PUBLISHER: Dan Savage

MARKETING MANAGER:
Charlotte Park

COMMERCIAL DIRECTOR:
Nigel Hole

PUBLISHED BY:
Mortons Media Group Ltd, Media
Centre, Morton Way, Horncastle,
Lincolnshire LN9 6JR. Tel. 01507 529529

PRINTED BY: William Gibbons and
Sons, Wolverhampton

ISBN: 978-1-909128-81-1

BRITISH DINOSAURS

The history of dinosaurs began in England with discoveries that were to have a profound effect on the rest of the world for years to come.

At the beginning of 19th century no one in Britain or anywhere else in the world had ever heard of dinosaurs. There were no monsters in history except perhaps those mentioned in the bible or within the pages of fanciful medieval bestiaries.

Few had even the faintest idea that the earth beneath their feet was already 4.5 billion years old when Jesus Christ was born. Even those who knew a little about geology had no real notion of what Britain might have been like millions of years ago.

The average person went about their difficult daily life without giving a thought to prehistory. But this all changed thanks to the Industrial Revolution and the arrival of the great age of steam power.

As factories were opened and expanded to build all the new iron and steel machines required by industrial Britain, more workers were needed. People flooded into the cities from the countryside and they had to have somewhere to live.

Huge housing projects were begun and demand for bricks and mortar rocketed. Existing quarries were working overtime and new quarries were opened up all over the country. With so many people digging into the earth and stone it was only a matter of time before they began to discover a multitude of what were unmistakably fossilised bones.

Large bones had been found before – for example in 1677 Oxford University chemistry professor Robert Plot described an enormous piece of thigh bone in his Natural History of Oxfordshire – but these had been explained as belonging to giant humans.

Now the brightest minds in science began to take an interest in these discoveries and to actively seek them out themselves. They quickly realised that these remains could not have belonged to anything human.

The breakthrough came during a meeting of the Geological Society of London on February 20, 1824, when William Buckland announced that a collection of fossils he had been studying for several years belonged to a giant lizard that he called Megalosaurus. This was the first dinosaur ever named.

A second dinosaur, Iguanodon, was discovered in 1822 and studied by Gideon Mantell, who gave it its name in 1825. He then identified a third species, Hylaeosaurus, in 1832.

It was not until 1842 that another palaeontologist, Richard Owen, came up with the term 'dinosaur'

Geologist William Buckland from Devon wrote the first scientific paper to describe a dinosaur – Megalosaurus – in 1824. This was the first dinosaur ever named. *Wellcome Trust*

or 'dinosauria'. Before that, the three species had been referred to simply as 'great fossil lizards'.

SPREADING THE WORD

By now these strange beasts, apparently unearthed by some of the most learned men in Britain, had begun to attract national attention. The newspapers carried reports of the great land leviathans being discovered and ordinary working people began to take notice.

Public enthusiasm increased – slowly at first but then picking up speed. The hunger for more information about the great animals that once roamed the earth began to grow rapidly. Soon amateur palaeontologists were scouring local cliff faces, beaches, quarries and brick-clay pits trying to find their own piece of prehistory.

Megalosaurus in particular really fired up the minds of the population as news of its terrifying teeth and massive size spread. In fact Megalosaurus was so publicly popular that it is even mentioned in the open line of Charles Dickens' 1853 novel Bleak House: "As much mud in the streets, as if the waters had but newly retired from the face of the earth, and it would not be wonderful to meet a Megalosaurus, forty feet long or so, waddling like an elephantine lizard up Holborn Hill."

That line also betrays the problems faced by early palaeontologists. This was a very new science and

XXI.—Ideal scene in the Lower Cretaceous Period, with Iguanodon and Megalosaurus.

Surgeon and fossil hunter Gideon Mantell from Sussex discovered Iguanodon in 1822 but did not write up his findings until 1825. When he died in 1852, he was still credited with discovering four of the five types of dinosaur then thought to have been discovered.

Victorians imagined Iguanodon and Megalosaurus as great sluggish lizards that may have fought one another in scenes such as this 1865 depiction by Edouard Riou.

the early identification of fossilised bones with little to compare them to and without modern equipment was not an easy task.

Simply identifying which bone was from which animal, let alone which part of an animal, was virtually impossible and this led to much heated debate and disagreement.

It is worth remembering that British dinosaur finds were very seldom the complete, polished specimens that we might see in museums today. Many were just single bones or a collection of fragments which made initial identification incredibly difficult. As a result, the early pioneers of fossil finds were incorrect in much of what they thought.

Nevertheless, in response to the huge level of interest, in 1852 huge concrete sculptures were commissioned for a display outside the Crystal Palace

following its move from Hyde Park to Bromley in south London. Owen directed sculptor Benjamin Waterhouse Hawkins on how he thought Megalosaurus and Iguanodon should look and the results were unveiled to great acclaim in 1854. These days it's easy to see how wrong these ideas of the early giants were, but at the time Hawkins' visions were a revelation.

SEEING THE FIRST 'DINOSAUR'

Imagine that there are no books on dinosaurs – that you have never even seen a picture of one. Imagine too, that you never went to school and that all you know is the world around you.

Then one day you visit a park in south London and there, right in front of you, looms a giant concrete statue. Rising out of some carefully-positioned plants,

Biologist and naturalist Richard Owen, from Lancaster in Lancashire, was implacably opposed to the idea of evolution but coined the term 'dinosaur' and discovered several different species. He also hated Gideon Mantell. Mantell in turn described him as "talented, dastardly and envious".

Dinosaur models under construction at Benjamin Waterhouse Hawkins' studio in Sydenham, circa 1853. They were to go on display outside the Crystal Palace the following year.

its massive shoulders puffed up on top of impossibly thick, tree trunk-type legs, its outer layer covered in skin as thick as armour and covered with hexagonal slabs of protective plates.

Its huge jaws and teeth barely contain a snarl that echoes the menace in its blood-red eyes, large and glaring from beneath a heavy, swollen brow ridge of muscle and bone. Bigger than any vehicle of the time, it looked like a cross between nature's fiercest predator and the stuff of nightmares.

Perhaps you dismiss the beast as some flight of fancy by the artist but no, you're soon informed that this thing is real – or was real; that it once roamed what is now England and hunted right where you are standing. Imagine finding that out for the first time.

And that was how the people of Victorian England came face to face with the first dinosaurs. British scientists informed the rest of the world of their discoveries and before long more dinosaurs were being unearthed all over the globe. England led the way with dinosaurs and we've not looked back since.

A pair of Iguanodon, Victorian style, as they appear in Crystal Palace Park. *Ian Wright*

THE LATEST TECHNOLOGY

While the original fossil hunters of the last century had to rely on interpretation and, to some degree, imagination to visualise and identify the animals they discovered, today's dinosaur hunters have a wealth of technology to use.

Electromagnetic, micron-level microscopes, hugely advanced computer programs and 3D virtual sculpting techniques are now commonplace in the quest to discover what these great animals were really like.

Medical grade scanners are also employed and by using their ability to look inside something without having to break it open we can learn so much more about the realities of dinosaur life.

Using a machine called a synchrotron, scientists can scan a fossilised egg and have already examined the unhatched, fossilised baby dinosaurs that have remained trapped in their shell for the last 70 million years.

From these scans we know what sort of skin these dinosaurs had, how their skeletons formed and even how many teeth they were born with. All this from a fossilised dinosaur in its shell that measures no more than a few inches across.

Modern technology can even help us restore fossils where only a fragment of the original remains. Dr Stephan Lautenschlager from Bristol University has pioneering a technique that uses a high-resolution X-ray and Computed Tomography (CT scanning) plus digital visualisation.

His team used the combination of advanced computer technology on the crushed and incomplete fossilised skull of an Erlikosaurus Andrewsi – a one-of-a-kind Therizinosaur. The fossil was scanned then digitally taken apart. The gaps in the bone structure were then filled in and it was rebuilt by the computer. Muscles, tendons and ligaments were then added to the picture before finally the skin was overlaid onto the reconstruction.

HOW DO WE KNOW HOW DINOSAURS MOVED?

One particularly fascinating area of dinosaur study involves ichnofossils or 'trace fossils'. This is essentially any mark, impression or other evidence that provides a geological record of how dinosaurs moved.

These can be scratches in the ground, dinosaur footprints or even dinosaur droppings – a host of which have survived remarkably well preserved, providing an instant snapshot of dinosaurs as they moved and lived.

The oldest types of tetrapod tail and foot prints come from the Devonian period and have been found in both Ireland and Scotland. Based on the position, depth and frequency of the foot imprints and the length and undulation of the tail dragging on the floor the prehistoric creature's height, leg length, gait and weight can be worked out.

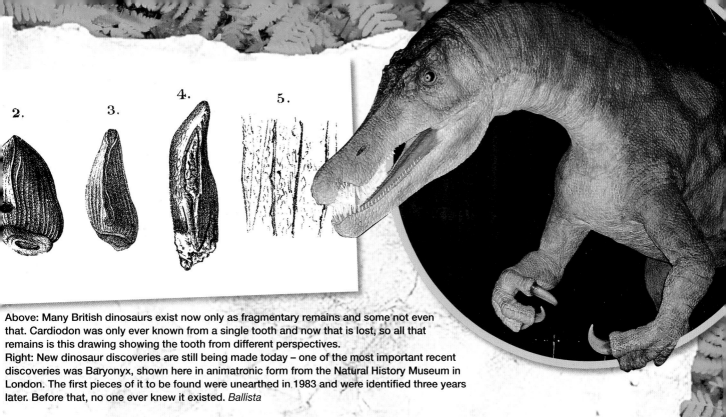

Above: Many British dinosaurs exist now only as fragmentary remains and some not even that. Cardiodon was only ever known from a single tooth and now that is lost, so all that remains is this drawing showing the tooth from different perspectives.
Right: New dinosaur discoveries are still being made today – one of the most important recent discoveries was Baryonyx, shown here in animatronic form from the Natural History Museum in London. The first pieces of it to be found were unearthed in 1983 and were identified three years later. Before that, no one ever knew it existed. *Ballista*

The new technology is very exciting – even allowing some palaeontologists to examine bones and remains of animals while they are still in the ground (although to paraphrase Dr Alan Grant from Jurassic Park, "Where's the fun in that?") – but you might wonder how detailed it can get. Well, how about this; a team in Germany recently digitally deconstructed and rebuilt the fossilised remains of the earliest known bird to pollinate plants Pumiliornis tessellatus. The fossil wasn't technically that of a dinosaur – being a mere 47 million years old instead of the 65 million years which is the age at which the 'great extinction' of dinosaurs happened – but the techniques used were sound and the team discovered that in the stomach of the early bird were pollen grains attached to the fruits it fed on.

3D modelling has also given us the first genuine idea of how a dinosaur truly sounded. After CT scanning the hollow crest of a Parasaurolophus, scientists at Sandia National Laboratories and the New Mexico Museum of Natural History and Science simulated what it might sound like if air was blown through it – a sort of booming, trumpeting noise.

While it is rare for a British dinosaur to be found in such a complete state that a full reconstruction is possible, new discoveries are being made every year which add to the picture of what these islands were like between around 220 and 65 million years ago when they were ruled by giant creatures the like of which we will never see again.

A surprising number of dinosaur footprints have survived into the modern age. This one is housed at the Hunterian Museum in Glasgow, Scotland, and is believed to be the world's smallest dinosaur footprint – the creature that made it was no bigger than a blackbird. *Osama Shukir Muhammed Amin*

How can you tell a dinosaur's colour?

It sounds like something from the realms of science fiction, but it is possible to tell, in some rare cases, whether a dinosaur was lightly or darkly coloured. These days we can even tell that some dinosaurs had feathers and how dark they were.

It was recently found that tiny microscopic elements in fossils, originally thought to be bacteria, are actually melanosomes, small elements within cells that contain melanin which is the organic compound that gives all living things with hair, feathers, skin and eyes their colour.

Different combinations of these tiny building blocks give different colours in animals.

The first fossil melanosomes were described in a fossil feather by Dr Jakob Vinther, a palaeobiologist at the University of Bristol in 2008. Since that discovery there have been literally thousands of finds giving us a much better idea of what shades dinosaur skin and feathers would have been.

A pair of Saltopus emerge from the forests of Triassic Britain – part of the supercontinent Pangaea.

DAWN OF THE DINOSAURS
THE TRIASSIC PERIOD
[251-205 MILLION YEARS AGO]

At the beginning of the Triassic period 251 million years ago, what we know today as Britain was a landlocked area to the north east of the gigantic supercontinent Pangaea.

What is now Scotland lay to the east of England, and in England itself great rivers flowed northwards and deposited large quantities of sand, grit and gravel over the areas that now lie between Nottingham and Darlington, and Birmingham and Carlisle.

The landscape would have looked very different to how it is today. It was rocky, sandy and barren, with no grass, just ferns and mosses at ground level and sparse trees with spikey leaves up above. There were no flowering plants so everything was either green or brown. While Britain had flowing water and plants, there is evidence that it also went through periods of severe drought when sand dunes formed and life would have become difficult for its inhabitants.

In spite of this, the creatures living here at the time could count themselves lucky. The earth had just gone through the greatest mass

A map of the British Isles showing in pink the layers of rock that survive from the Triassic period.

possibly a combination of these over an extended period of time, no one knows for sure.

As a result though, the Early Triassic was generally sparsely populated and the earth no longer had its characteristic abundant diversity of life. The only things that seemed to thrive during this period were fungi that fed upon the remains of the dead.

It took about 10 million years for ecosystems to begin recovering and it was not until 20 million years later that larger animals began to make a comeback.

During the Late Triassic, another extinction wiped out many pre-dinosaur creatures – leaving gaps in the food chain which the first dinosaurs then managed to fill. In addition, there were no mammals, no birds and no lizards, only lizard-like reptiles, insects such as dragonflies, millipedes and scorpions. Dinosaurs were the only large land animals on earth at this point.

By now, the sea had begun to encroach as Pangaea broke up and Britain became first a coastal peninsula on the eastern side of the continent and then an archipelago – small islands that had previously been areas of desert upland. The direction of continental movement also meant that Scotland had shifted position to the north west of what is now England.

The temperature was generally warm and subtropical, the seas were shallow and vegetation began to spread and grow rapidly across the areas of dry land. It was into this environment that the earliest British dinosaurs – including first Saltopus and then Agnosphitys, Camelotia, Gresslyosaurus and Zanclodon – emerged.

What is now England (the red dot) and Scotland (the blue dot) as they were positioned 250 million years ago – before the dinosaurs.

extinction event in history, with 90% of all species of plants and animals being wiped out. What forests there had been disappeared.

Exactly what caused this is unknown. Whether it was a massive meteor strike, a sudden release of toxic gases from the sea, the sea itself becoming poisoned by hydrogen sulphide, extreme volcanic activity or

The sea begins to encroach on Pangaea during the Mid-Triassic, towards the landlocked area that is now part of Britain.

By the Late Triassic, the positions of England and Scotland had shifted dramatically.

SALTOPUS
HOPPING FOOT

About the size of a domestic modern cat and actually built more like a featherless modern chicken, Saltopus is considered to be Scotland's and probably Britain's oldest dinosaur. It was a quick, bird-like meat-eater that was very small and had hollow bones. Most of its 1m length was made up of its light, slender tail. Unusually for a dinosaur, Saltopus had five fingers on each hand although both the fourth and fifth fingers on each were reduced in size. In dinosaurs that appeared after Saltopus the fourth and fifth fingers had disappeared.

This small dinosaur was found in the Lossiemouth Sandstone Formation in Scotland, so it can be dated from the Carnian-Norian stage of the Late Triassic period.

At just 1kg in total weight this would have been a light-footed, fast-moving animal that darted about very much like big modern birds do today and it is thought that it would have been active and frantic in movement, seldom staying still for long.

It would have preyed upon anything smaller that it could catch with its long mouth and small, sharp teeth. Insects and scavenged carcass meat would have made up a serious proportion of this dinosaur's diet.

VITAL STATISTICS

NAME: Saltopus (sawll-toe-puss)

NAME MEANING: Hopping Foot

FAMILY: Silesauridae

ESTIMATED SIZE: 1m long

ESTIMATED WEIGHT: 1kg

DIET: Carnivore

ANATOMICAL CHARACTERISTICS: Small and bipedal with sharp teeth

LIVED DURING: Carnian-Norian stage of the Late Triassic

WHERE FOUND: Lossiemouth West and East Quarries, Scotland

FOSSIL REMAINS: Spinal column, forelimbs, pelvis and legs

AGNOSPHITYS
UNKNOWN BEGETTER

One of Britain's earliest dinosaurs or dinosaur-like reptiles, like Saltopus, little Agnosphitys existed in a time before colossal land animals.

The area where it lived, Somerset, was a desert during the Mid-Triassic. Warm, shallow seas lapped against sandy beaches in the area, turning into a broad desert upland with the rapidly eroding remains of a huge mountain range – of which today's Mendips were once part. Walking around on two legs, Agnosphitys would have hunted for insects or small reptiles among the sparse vegetation. It may also have eaten parts of that vegetation to supplement its diet.

Towards the end of the Triassic, the sea began to encroach more and more on Agnosphitys' habitat until, by the end of the period, Somerset had been transformed into a rocky archipelago. Deep fissures opened up in the rocks and periodic flash flooding swept animals such as Agnosphitys into them. It was in one of these fissures that the only known remains of Agnosphitys were discovered during the 1990s – in what today is known as Slickstones Quarry.

It was named Agnosphitys or 'Unknown Begetter' because no-one is quite sure whether it really is a dinosaur or just a very dinosaur-like reptile. If it was a dinosaur, it was a carnivore, if it was a reptile it may have been an omnivore; it certainly had pointed and serrated teeth. The 'begetter' part of its name comes from the fact that Agnosphitys may have been an ancestor to the much larger dinosaurs that came later.

VITAL STATISTICS

NAME: Agnosphitys (ag-nohs-FIE-tis)

NAME MEANING: Unknown Begetter

FAMILY: Guaibasauridae

ESTIMATED SIZE: 0.7m long

ESTIMATED WEIGHT: 4kg

DIET: Carnivore or omnivore

ANATOMICAL CHARACTERISTICS: Small and bipedal with sharp teeth

LIVED DURING: Norian stage of the Late Triassic

WHERE FOUND: Slickstones (Cromhall) Quarry, South Gloucestershire, England

FOSSIL REMAINS: Left upper jawbone, left and right ankle joints, right funny bone, one tooth

CAMELOTIA
FROM CAMELOT

Camelotia lived in the same region as Agnosphitys but at a time when the smaller animal was in decline or had already become extinct.

Dense vegetation had replaced the rocky desert of earlier times and Camelotia was adapted to take full advantage of this. It walked on all fours but had longer hind legs, enabling it to stand up against a tree and use its long neck to reach the leaves high above.

At between nine and 11 metres long and weighing between two-and-a-half and three tonnes, Camelotia was probably the largest animal that had ever lived in Britain up to that point. This alone may have been enough to deter the predators of that era, potentially including the likes of Zanclodon.

Remains of several meat-eating dinosaurs were found jumbled up with those of Camelotia and it has been suggested that one particular tooth might represent evidence of something feeding on a Camelotia's remains.

Most of what has been identified as Camelotia was discovered at Wedmore Hill in Somerset, but a tooth and a claw found at Staple Pit, near Newark in Nottinghamshire, may also have belonged to the species – giving it a fairly wide geographical spread.

Camelotia was first named in the 1890s as Avalonia sandfordi but sometime later it was found that this name had already been used for something else, so its name was changed to Avalonianus. A further review of the Avalonianus fossils in 1985 revealed that some of them didn't actually belong to the same species at all, so finally what could be identified as definitely belonging to one animal type was separated out and given the name Camelotia.

VITAL STATISTICS

NAME: Camelotia (kam-eh-loh-tee-uh)

NAME MEANING: From Camelot

FAMILY: Melanorosauridae

ESTIMATED SIZE: 9-11m long

ESTIMATED WEIGHT: 2.5-3 tonnes

DIET: Herbivore

ANATOMICAL CHARACTERISTICS:
Long neck, long tail

LIVED DURING: Rhaetian stage of the Late
Triassic to Early Jurassic

WHERE FOUND: Westbury Formation,
Wedmore, Somerset, England

FOSSIL REMAINS: Backbone, hip bones
and hind leg bones

GRESSLYOSAURUS
GRESSLY'S LIZARD

Living at a time when Britain was warm and gradually becoming covered with vegetation, Gresslyosaurus walked around on its hind legs and used its muscular arms to grip its food and direct it towards its small head. Inside its mouth were sharp but thick teeth ideal for crushing plant matter. It lived in herds and adults could reach an unusually wide range of sizes, from 4.8m to 10m long, weighing between 600kg and four tonnes. It is thought that Gresslyosaurus lived for between 12 and 20 years.

Very little of what might be identified as Gresslyosaurus has been found in Britain but nevertheless there are fragments which could indicate that the species, closely associated with the better known Plateosaurus, lived in more northerly areas. The species, named after Swiss palaeontologist Amanz Gressly, was discovered in the 1800s in Switzerland and has since been grouped together with Plateosaurus and Plateosaurus-type specimens from Switzerland and Germany.

In 1997, engineers drilling for oil at the Snorre offshore field in the North Sea found the crushed knucklebone of what turned out to be a large dinosaur more than 1.4 miles below the seabed. After much careful study, in 2006 this was identified as belonging to a Plateosaurus.

Some 210 million years ago, that part of the North Sea was a vast dry sandy plain crossed by occasional rivers and it has been surmised that Plateosaurus, and species like it, migrated across huge areas between what is now southern Europe and Greenland – including the northern parts of Britain.

This would make Gresslyosaurus a visitor to Britain, rather than an exclusively British dinosaur.

VITAL STATISTICS

NAME: Gresslyosaurus
(gress-lee-oh-sore-us)

NAME MEANING: Gressly's Lizard

FAMILY: Plateosauridae

ESTIMATED SIZE: 4.8-10m long

ESTIMATED WEIGHT: 600kg to 4 tonnes

DIET: Herbivore

ANATOMICAL CHARACTERISTICS: Small head on a long mobile neck, powerful arms with grasping claws

LIVED DURING: Middle Norian to Rhaetian stage of the Late Triassic

WHERE FOUND: Staple Pit, near Newark, Nottinghamshire, England

FOSSIL REMAINS: Fragmented bits and pieces

ZANCLODON
SCYTHE TOOTH

The most fearsome looking predator in Britain during the Late Triassic may well have been Zanclodon. As plant-eaters grew in size, so too did the meat-eaters that preyed upon them.

Zanclodon was a carnivore boasting long jaws lined with pointed recurved teeth. Its size would have enabled it to tackle smaller plant-eaters such as Gresslyosaurus but the largest herbivores, such as Camelotia, may have been beyond it.

Unlike later carnivores, given its size Zanclodon may have been relatively slow and lumbering – relying on the weaknesses of its prey to make an easy kill.

Alternatively, it may have dined primarily on the bodies of animals that had already died.

The lower jawbone impression from which Zanclodon is known was found at Stormy Down near Bridgend, South Wales, in 1898. In recent years it has been suggested that the fossil may in fact belong to a crocodile-like reptile.

VITAL STATISTICS

NAME: Zanclodon

NAME MEANING: Scythe Tooth

FAMILY: Teratosauridae

ESTIMATED SIZE: 4m long

ESTIMATED WEIGHT: 500kg

DIET: Carnivore

ANATOMICAL CHARACTERISTICS: Large and bipedal with short arms, thick tail and sharp teeth

LIVED DURING: Rhaetian stage of the Late Triassic and Early Jurassic

WHERE FOUND: Stormy Down, Kenfig, near Bridgend, Glamorganshire, Wales

FOSSIL REMAINS: Several teeth and impression of lower jawbone

THECODONTOSAURUS
SOCKET-TOOTHED LIZARD

Thecodontosaurus was a little plant-eating dinosaur that had a small head and shorter forelimbs than legs. Moving on two legs, its long tail would have helped balance the body, neck and head as it broke into a run. It would have needed to be able to get a move on if it wanted to avoid predators. One notable thing about its five-fingered hands was that the thumb was extended with a large, curved claw. It is unknown exactly what this would have been for but it was most likely used to help Thecodontosaurus get at its food.

Unsurprisingly it was named after its very distinctive teeth which are similar to those found in a modern monitor lizard but rather than being secured into its gums by roots, these were anchored firmly into sockets in the jaws.

All the best-preserved fossils of Thecodontosaurus were destroyed during the Second World War. In November 1940, German bombs landed on and devastated the geology gallery of Bristol City Museum where they were on display.

Some of the surviving remains once thought to have belonged to Thecodontosaurus have since been discovered to be those of a different species – Pantydraco.

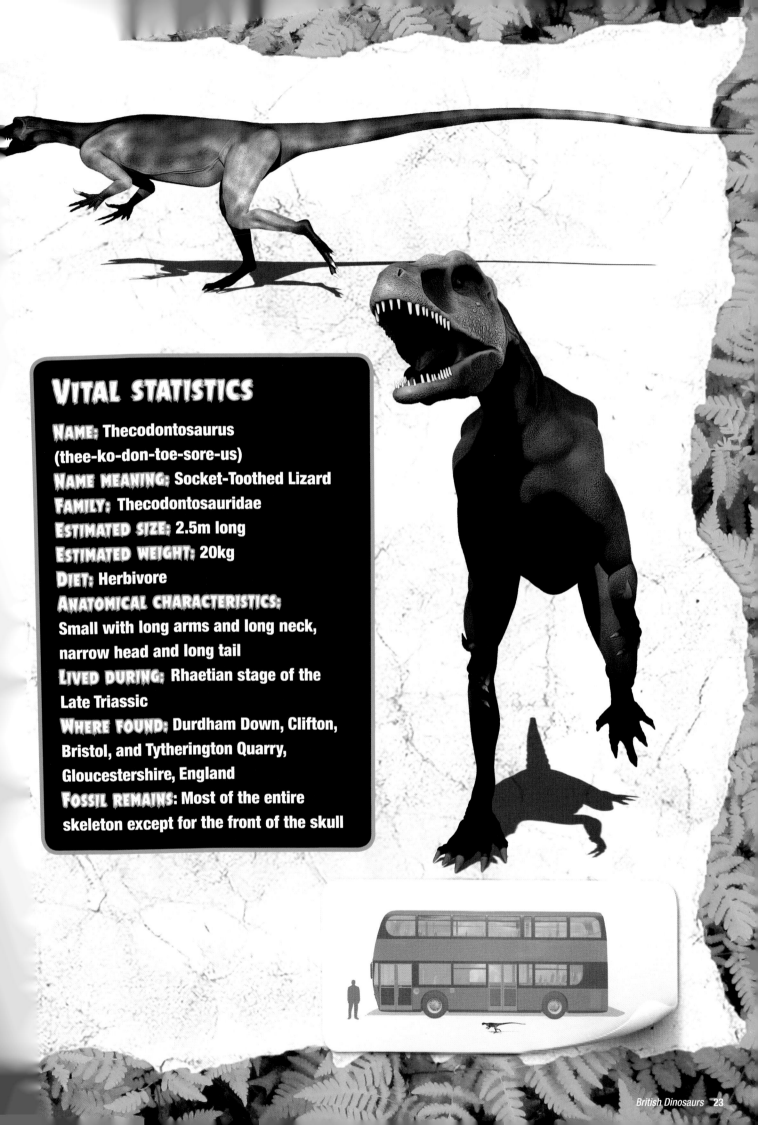

VITAL STATISTICS

NAME: Thecodontosaurus
(thee-ko-don-toe-sore-us)
NAME MEANING: Socket-Toothed Lizard
FAMILY: Thecodontosauridae
ESTIMATED SIZE: 2.5m long
ESTIMATED WEIGHT: 20kg
DIET: Herbivore
ANATOMICAL CHARACTERISTICS:
Small with long arms and long neck,
narrow head and long tail
LIVED DURING: Rhaetian stage of the
Late Triassic
WHERE FOUND: Durdham Down, Clifton,
Bristol, and Tytherington Quarry,
Gloucestershire, England
FOSSIL REMAINS: Most of the entire
skeleton except for the front of the skull

DINOSAUR ISLANDS
THE JURASSIC PERIOD
[205-142 MILLION YEARS AGO]

The warm shallow sea that flowed between the scattered islands of Britain at the start of the Jurassic period brought with it large amounts of mud and a wide diversity of marine life.

The ammonite Psiloceras – the only surviving ammonite after the mass extinction at the beginning of the Triassic – began to arrive in the waters around Britain along with a hard-shelled squid-like creature called Passaloteuthis.

Molluscs like the 'Devil's Toenail' Gryphaea, brachiopods, crinoids and fish appeared too. Following after them came large fish-eating marine reptiles such as ichthyosaurs and plesiosaurs. Temperatures throughout the year ranged from a brisk but not freezing 12°C in the depths of winter to a sweltering 29°C during the height of summer.

By the time of the Mid-Jurassic, while most of what is now southern England and Wales was the watery domain of these creatures, parts of northern England, Scotland and East Anglia were becoming substantial islands covered with forests.

These would have looked a little more like the forests of today than the sparse and odd-looking plant cover that had been present during the Triassic. Conifer trees loomed overhead, with ginkgos and horsetails both growing tall. Down below were ferns, slender horsetails and tough woody cycads.

Living among these were reptiles, small primitive mammals and dinosaurs. There were rivers and streams too, populated by clams, snails and fish. Areas of the East Midlands, North Yorkshire and Western Scotland developed into lagoon or river delta type environments.

During the Late Jurassic, sea levels began to fall, the continental mass began to rise and large areas of southern England emerged from the ocean. Every now and then a period of low oxygen in the water caused the deaths of many of the large marine reptiles.

These had changed too – by the Late Jurassic ichthyosaurs more closely resembled modern dolphins with dorsal fins and tail flukes. Enormous short-necked pliosaurs such as Liopleurodon 'Smooth-Sided Teeth' were the dominant ocean-going predators boasting long powerful jaws filled with big sharp teeth.

The rocks bearing Jurassic era fossils are highlighted in blue on this map of Britain.

Britain was a collection of small islands at the beginning of the Jurassic period (above left) stalked by a wide range of dinosaurs.

Over land and sea flew pterosaurs. The Early Jurassic saw Dimorphodon take flight. It was a 1m long, 1.45m wingspan, meat-eater with a large sharp-toothed skull that looked more like it belonged on a ground-dwelling carnivore. Later came the 2m wingspan Rhamphocephalus and others.

Jurassic Britain was home to the most diverse range of species yet seen, with reptiles ruling the seas and skies while dinosaurs developed rapidly on the British islands themselves.

Dinosaurs had started out relatively small during the Triassic but by now they were growing quickly in size. The plant-eaters grew to take advantage of the lush foliage now available – including flowering plants for the first time but still no grass. The meat-eaters of the period also got bigger to take advantage of the larger quantities of food that were on offer.

Throughout the Jurassic all these animals continued to evolve, ultimately leading to the more specialised dinosaurs of the Cretaceous period.

The small islands that made up Britain became more substantial during the Mid-Jurassic.

By the Late Jurassic, the British islands had become linked to mainland Europe.

PANTYDRACO
PANT-Y-FFYNNON DRAGON

Pantydraco was an omnivorous dinosaur which in the early part of the Jurassic period would hunt and graze in the west woodlands that made up what is now known as France, the south of England and southern Wales. It would have been fast and bird-like with its movements on the ground, active and clever enough to find food across a large expanse of land.

In terms of the evolution of dinosaurs, Pantydraco is an important specimen. That's because it appeared right between the meat-eaters of its ancestors and the herbivores of its descendants. As such, there is some debate over exactly what would have made up Pantydraco's diet.

Although skeletal finds are often represented as being bipedal it seems entirely possible (especially given how this dinosaur was between two phases of dinosaur evolution) that Pantydraco was able to walk on either two or four legs, depending on what it was doing.

Three fingers on the forearms would have been both useful for grabbing food but given the spread of the hands of this type of animal, would also have been useful as a second pair of feet when walking.

VITAL STATISTICS

NAME: Pantydraco (pant-uh-dray-ko)

NAME MEANING: Pant-y-ffynnon Dragon

FAMILY: Saurischia

ESTIMATED SIZE: 3m long

ESTIMATED WEIGHT: 65kg

DIET: Omnivore

ANATOMICAL CHARACTERISTICS: Small and bipedal with sharp teeth

LIVED DURING: Rhaetian stage of the Late Triassic and into the Early Jurassic

WHERE FOUND: Pant-y-ffynnon quarry, Bonvilston, South Glamorgan, South Wales

FOSSIL REMAINS: Skull and partial skeleton including neck, partial shoulder girdle, forelimb bones

Sarcosaurus
Flesh Lizard

Sarcosaurs were meat-eaters with a very aggressive look which was typical of some of the carnivores of the era. What wasn't typical of theropods at the time Sarcosaurus roamed was an animal of this type and this size. For the Sineumurian stage of the Early Jurassic period this was a very large predator indeed. It ruled the western woodlands of what is now Europe while it hunted.

Add that to the sharp teeth set at the front of the mouth, the elongated neck that was roughly equal to the comparatively thick tail and the long forearms and you have what amounts to a formidable predator. It feasted on smaller dinosaurs, large early insects and any carrion it came across.

Sarcosaurus is thought to have also hunted along the edge of rivers and lakes. Certainly, if it encountered any difficulty in catching particular prey then this dinosaur would have turned its attention to whatever alternative it could get its powerful jaws around – including early fish and other water-based animals of the time.

Originally named in 1921 by Charles William Andrews, Sarcosaurus remains have been mixed up with those of both Magnosaurus and Megalosaurus over the years since.

VITAL STATISTICS

NAME: Sarcosaurus (sarr-ko-sore-us)

NAME MEANING: Flesh Lizard

FAMILY: Coelophysoidea

ESTIMATED SIZE: 3.5m long

ESTIMATED WEIGHT: 50kg

DIET: Carnivore

ANATOMICAL CHARACTERISTICS:
Bipedal with sharp teeth

LIVED DURING: Sinemurian stage of the
Early Jurassic

WHERE FOUND: Barrow upon Soar,
Leicestershire, England

FOSSIL REMAINS: Pelvis, part of the spine
and the upper part of a leg bone

SCELIDOSAURUS
LIMB LIZARD

This is one of the most important dinosaurs ever to come out of Britain because the fossils of it found are so complete they have told us a lot about the animals of the time.

Scelidosaurus lived around 191 million years ago and grazed on low-level hard vegetation which took a lot of chewing before it was broken down to the goodness and useful nutrients needed by the dinosaur. Outwardly, this animal looked not dissimilar to other ankylosaurs – indeed this is one of the oldest known and most primitive members of the family.

Part of that familiar look is due to the armoured areas that covered most of the back of the animal, from the base of its skull all the way to the tip of its tail. Scelidosaurus was protected from attack by long horizontal rows of rolled over thickened skin plates that looked knobbly but were tough to get through.

Although this dinosaur was nearly 4m long, its head was just 20cm from base of skull to tip of jaw. The head wasn't high if viewed from the side but was longer than it was wide and from the top it looked slightly triangular.

Scelidosaurus had a snout that sat on top of the skull and the jaw was served by very strong muscles able to crush their way through the fibrous material that made up the majority of the dinosaur's diet.

VITAL STATISTICS

NAME: Scelidosaurus
(skell-ee-doe-sore-us)

NAME MEANING: Limb Lizard

FAMILY: Scelidosauridae

ESTIMATED SIZE: 3.8m long

ESTIMATED WEIGHT: 270kg

DIET: Herbivore

ANATOMICAL CHARACTERISTICS:
Walking on all fours with back legs
longer than front ones

LIVED DURING: Sinemurian to
Pliensbachian stages of the
Early Jurassic

WHERE FOUND: Black Ven cliffs,
Charmouth, Dorset, England

FOSSIL REMAINS: Complete specimen

MAGNOSAURUS
LARGE LIZARD

Magnosaurus was a large dinosaur that attacked and ate smaller, slower prey during the Mid-Jurassic era. It was a prime meat-eater of the time and had to be able to get across the large collection of land-linked islands that made up Britain at this time.

Most of Northern Europe during the Mid-Jurassic period was covered in warm seas but the land that was left above the water line was dense forest filled with smaller dinosaurs and other animals and these would have provided much of the Magnosaurus' diet. In order to be able to eat, this dinosaur had to be speedy and have good eyesight and other senses so that it could hunt its prey.

Once thought to be a type of Megalosaur, Magnosaurus did share a lot of similar physical qualities with its larger relative. A sharp-toothed biped with a thick tail and strong neck, the adults were over 4m long and likely to have weighed up to around half a tonne.

Stooped forward in the typical predatory stance, Magnosaurus kept its head low as it scoured the land for anything it could eat.

Magnosaurus was first discovered in the 19th century by James Parker near Nethercombe in Dorset. Being a member of the Tetanurae family (meaning 'stiff tails' which includes all theropod dinosaurs including birds) puts the Magnosaurus in an impressive group of dinosaurs, and at around 175 million years old this is the oldest known member of the family.

VITAL STATISTICS

NAME: Magnosaurus (mag-no-sore-us)

NAME MEANING: Large Lizard

FAMILY: Megalosauridea

ESTIMATED SIZE: More than 4m long

ESTIMATED WEIGHT: 220kg

DIET: Carnivore

ANATOMICAL CHARACTERISTICS: Large bipedal predator with sharp teeth

LIVED DURING: Lower Bajocian stage of the Mid-Jurassic

WHERE FOUND: Nethercombe, north of Sherborne, Dorset, England

FOSSIL REMAINS: Partial skull, spine, partial right pubis and lower jaw, thought to be from a juvenile

CETIOSAURUS
WHALE LIZARD

With its huge body housing a gut ideally suited to processing tons of plant matter and a neck long enough to get its little head up into the tops of trees, Cetiosaurus was well adapted to its environment.

It lived on the islands and dense forests of what is now southern to central England. Walking on all four legs it travelled in herds – particularly along the coastlines of the shallow sea or close to inland rivers.

Cetiosaurus was one of the earliest dinosaurs to be found and then described – in 1825. It was named by Sir Richard Owen in 1842 – the same year that he coined the term 'dinosaur'. With little to compare it against, he thought its backbone looked a bit like that of a whale. He also thought, overall, that it would have looked like an enormous crocodile.

While Cetiosaurus as we know it today might, at first glance, look much like any other long-necked, long-tailed herbivorous dinosaur, that backbone Owen examined so closely was actually more primitive than those of many later species. Its vertebrae were thick and coarse whereas later animals of a similar size and shape had hollowed out vertebrae to save weight. This makes Cetiosaurus one of the earliest animals of its kind.

VITAL STATISTICS

NAME: Cetiosaurus (see-tee-oh-sore-us)

NAME MEANING: Whale Lizard

FAMILY: Cetiosauridae

ESTIMATED SIZE: 18-20m long

ESTIMATED WEIGHT: 11 tonnes

DIET: Herbivore

ANATOMICAL CHARACTERISTICS: Typical sauropod features – long neck, small head, large body, long tail, but with primitive backbone

LIVED DURING: Bajocian to Bathonian stage of the Mid-Jurassic

WHERE FOUND: Forest Marble Formation, several sites in the Bletchingdon area, Oxfordshire; Bajocian Rutland Formation, Williamson Cliffe Brickworks Quarry, Great Casterton, Rutland, and possibly sites in Northamptonshire and Gloucestershire, England

FOSSIL REMAINS: Partial skeletons of at least three individuals including adult and juvenile animals

CARDIODON
HEART TOOTH

The collection of forested islands that made up what is now the south-west of England during the Mid-Jurassic period were home to some of the first truly gigantic dinosaurs.

During this time the earliest sauropods – animals with small heads, long necks and tails, barrel-like bodies and thick pillar-like legs – appeared to browse the increasingly lush foliage that continued to sprout up.

Cardiodon was among a variety of similar species to flourish at this time, alongside the likes of Cetiosaurus. Their very size tended to help protect them from predators and enabled them to reach areas of greenery that other animals simply could not.

Unfortunately, so little remains of Cardiodon that it is not possible to be much more specific about it. A single tooth was discovered by Wiltshire naturalist Joseph Chaning Pearce during the 1830s or early 1840s and the species to which it belonged was given the name Cardiodon because it was somewhat heart-shaped.

Later, and for more than a century, it was thought that the Cardiodon tooth actually probably belonged to a Cetiosaurus anyway. Since 2003 however, detailed examination of the few surviving drawings of Pearce's tooth (the actual tooth having been lost along the way) and comparisons with teeth that definitely belonged to Cetiosaurus have determined that Cardiodon was a distinct and separate animal all along.

VITAL STATISTICS

NAME: Cardiodon (car-dee-oh-don)

NAME MEANING: Heart Tooth

FAMILY: Possibly Cetiosauridae or Turiasuria

ESTIMATED SIZE: 10-18m long

ESTIMATED WEIGHT: 5-15 tonnes

DIET: Herbivore

ANATOMICAL CHARACTERISTICS:
Too little known

LIVED DURING: Bathonian stage of the Mid-Jurassic

WHERE FOUND: Forest Marble Formation, Bradford on Avon, Wiltshire, England

FOSSIL REMAINS: A single tooth, now lost

MEGALOSAURUS
BIG LIZARD

Fearsome Megalosaurus was the apex predator of its habitat – sitting at the top of the food chain. Once a Megalosaur had grown into the full size and strength of adulthood (for the era in which it lived – later apex predators were bigger. Tyrannosaurus Rex, for example measured 12.5m long compared to the Megalosaurus' 7m) almost no other animal of the Jurassic period could bother it.

As a result these dinosaurs were able to roam freely across the British islands hunting and eating stegosaurs and sauropods. It was a bipedal dinosaur that held its body horizontal and used its thick tail to balance the weight of its large head filled with deadly teeth set into an exceptionally strong jaw. Megalosaur's powerful back legs gave it the speed necessary to bring down fast and agile prey too – there have been examples of flying reptile remains found alongside fossils of Megalosaurs. Megalosaurus had to have a strong upper body and neck too in order to force its way through thick areas of vegetation and attack prey animals.

This is a dinosaur that fits the mould of what most people would describe if you asked them to pick a favourite. Large headed, intelligent and often operating in a pack, Megalosaurs were the first dinosaurs to ever be named and studied by science.

There is nothing to suggest that Megalosaurs lived in family groups or herds, but in order to overcome their prey they may well have been pack, rather than solitary, hunters. A degree of societal interaction must therefore have taken place between these fiercely single-minded animals.

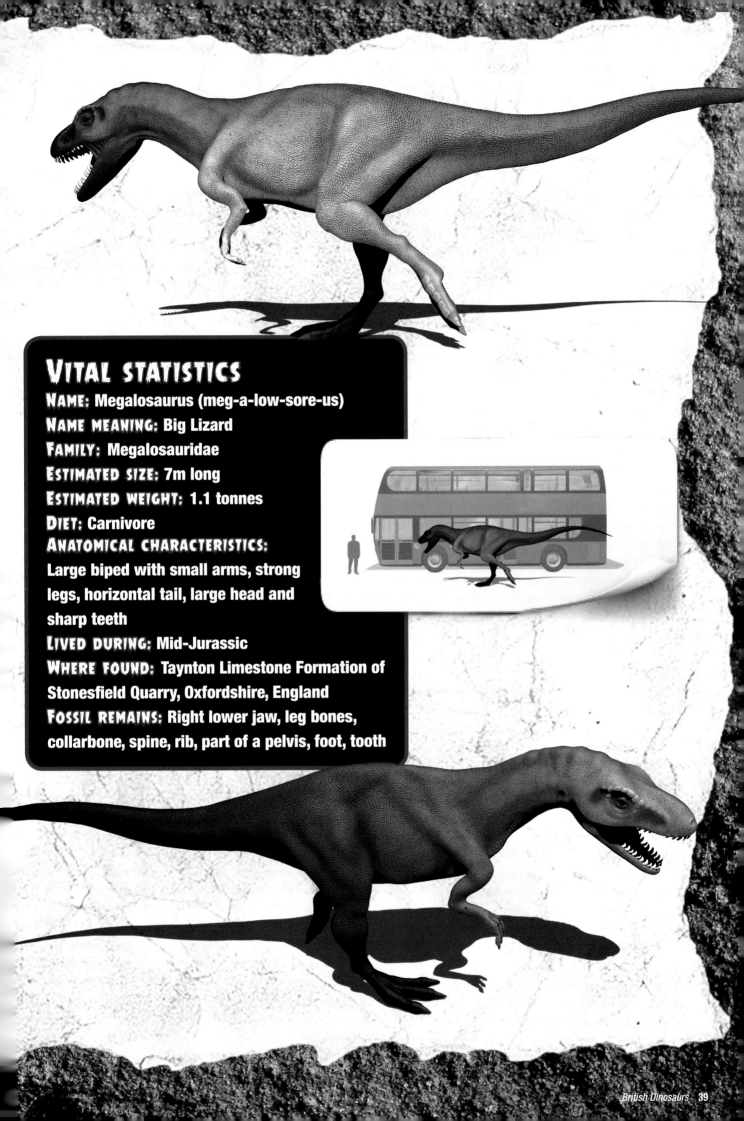

VITAL STATISTICS

NAME: Megalosaurus (meg-a-low-sore-us)

NAME MEANING: Big Lizard

FAMILY: Megalosauridae

ESTIMATED SIZE: 7m long

ESTIMATED WEIGHT: 1.1 tonnes

DIET: Carnivore

ANATOMICAL CHARACTERISTICS:
Large biped with small arms, strong legs, horizontal tail, large head and sharp teeth

LIVED DURING: Mid-Jurassic

WHERE FOUND: Taynton Limestone Formation of Stonesfield Quarry, Oxfordshire, England

FOSSIL REMAINS: Right lower jaw, leg bones, collarbone, spine, rib, part of a pelvis, foot, tooth

ILIOSUCHUS
CROCODILE HIPPED

Giants may have been walking the earth in the form of Cetiosaurus, Cardiodon and Megalosaurus but down below their feet – and perhaps beneath their notice – smaller creatures were locked in their own fight for survival.

One of these was a small carnivore called Iliosuchus. This unusual little dinosaur had a mouth full of sharp teeth and perhaps a coat of primitive feathers. It may have eaten lizards and other small animals or it might have scavenged from the carcasses of larger dinosaurs.

Running about on two legs, it was barely taller at the shoulder than the knees of an adult human today.

While it may have been an offshoot of the same family that gave rise to Megalosaurus, it is also possible the Iliosuchus is the earliest known member of the Tyrannosaurid family. This would make it an ancestor of the famous meat-eaters best known through Tyrannosaurus Rex.

The purpose of its coat of feathers, if it had one as many other small meat-eating dinosaurs did, is unclear. It may have been useful for insulation or it might have evolved for mating displays. Some scientists have suggested that dinosaurs like Iliosuchus had complex skins and may have been born with feather-like filaments which they then lost when they reached adulthood – although this tends to apply to larger dinosaurs. Too little remains of Iliosuchus to be sure.

VITAL STATISTICS

NAME: Iliosuchus (ill-ee-oh-soo-kus)

NAME MEANING: Crocodile Hipped

FAMILY: Tetanurae

ESTIMATED SIZE: 1.5m long

ESTIMATED WEIGHT: 50kg

DIET: Carnivore

ANATOMICAL CHARACTERISTICS: Sharp teeth and possible a coat of primitive feathers

LIVED DURING: Bathonian stage of the Mid-Jurassic

WHERE FOUND: Stonesfield slate mine, Spratt's Barn, Stonesfield, Oxfordshire, England

FOSSIL REMAINS: Three small hip bones and possibly a tooth

PROCERATOSAURUS
BEFORE HORNED LIZARD

While British rocks may not (so far) have yielded any Tyrannosaurus Rex remains, they have given up the near-perfect skull of a Proceratosaurus, a smaller relation of that most famous of dinosaurs.

The 30cm-long skull was recently re-examined and during this process it was determined that it belonged to one of the oldest known relatives of T Rex, another even smaller one of these being Iliosuchus.

The skull has spaces which show Proceratosaurus had powerful jaw muscles and its teeth, although smaller with than those of its fearsome relation, are banana-shaped and bear a very close resemblance to those of T Rex.

Proceratosaurus had an elaborate cranial crest and was identified easily by its nasal horn. It would have lived a largely solitary life apart from when it paired with a member of the opposite sex to breed. Fast moving and clever, this was an animal that spent most of its time hunting smaller animals in the woodlands southern England.

VITAL STATISTICS

NAME: Proceratosaurus
(pro-kerra-toe-sore-us)

NAME MEANING: Before
Horned Lizard

FAMILY: Proceratosauridea

ESTIMATED SIZE: 3m long

ESTIMATED WEIGHT: 60kg

DIET: Carnivore

**ANATOMICAL
CHARACTERISTICS:** Small and bipedal with
sharp teeth, similar to Tyrannosaurus Rex

LIVED DURING: Late Bathonian stage of
the Mid-Jurassic

WHERE FOUND: Minchinhampton,
Gloucestershire, England

FOSSIL REMAINS: Skull

CALLOVOSAURUS
CALLOVIAN LIZARD

Forests of enormous ferns, cycads and conifers covered the island that is now a large part of Cambridgeshire during the Mid-Jurassic and among those plants lived Callovosaurus.

While much larger animals such as Cetiosauriscus browsed among the upper foliage, the 1-1.5m tall Callovosaurus found its food among the smaller plants and shrubs.

Walking mostly on its hind legs but sometimes on all fours, in appearance it was probably something like an Iguanodon but much smaller and more delicate with a slightly longer body and a longer neck.

It would have had little difficulty finding food but it would also have found itself on the menu for some of the large meat-eaters living at around the same time, including Eustreptospondylus. Whether it would have had the speed and agility to out-manoeuvre or out-run these predators is difficult to tell from the very scant remains available for Callovosaurus, however.

First identified in 1889, the species was named after the time period when it lived – the Callovian stage of the Mid-Jurassic, between 166 and 163 million years ago. However, the Callovian period itself is named after a small hamlet near Chippenham in Wiltshire called Kellaways. Callovosaurus could therefore be regarded as being Kellaways Lizard.

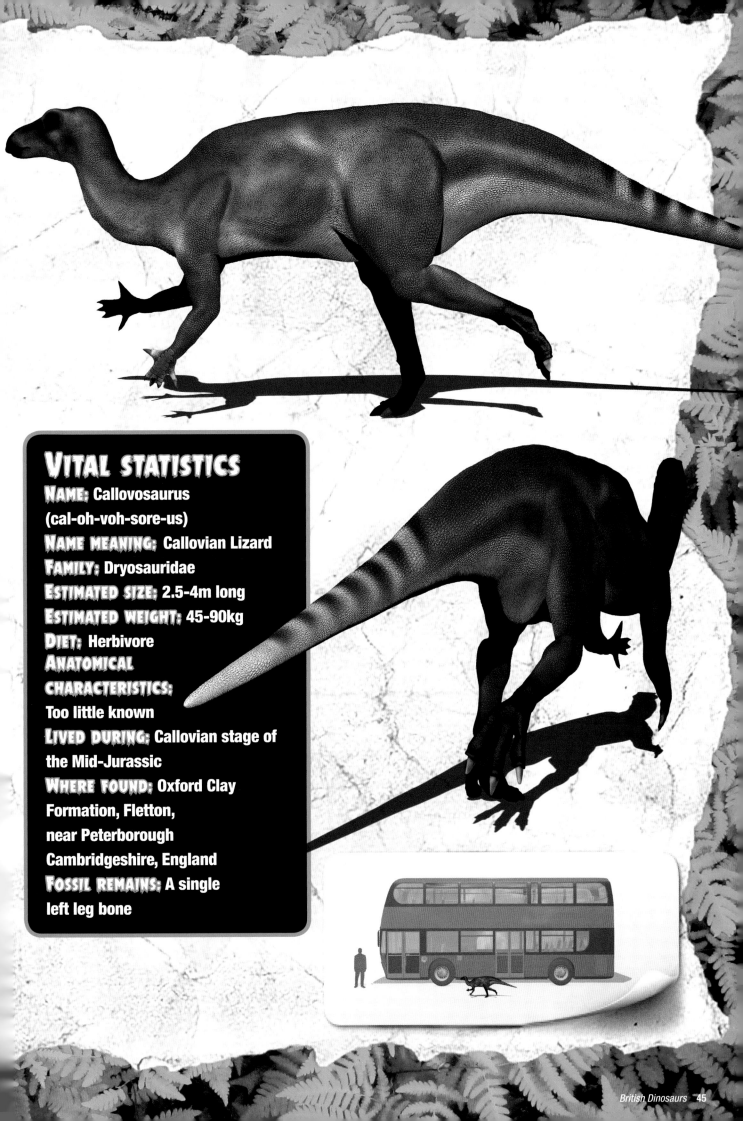

VITAL STATISTICS

NAME: Callovosaurus
(cal-oh-voh-sore-us)

NAME MEANING: Callovian Lizard

FAMILY: Dryosauridae

ESTIMATED SIZE: 2.5-4m long

ESTIMATED WEIGHT: 45-90kg

DIET: Herbivore

ANATOMICAL CHARACTERISTICS:
Too little known

LIVED DURING: Callovian stage of
the Mid-Jurassic

WHERE FOUND: Oxford Clay
Formation, Fletton,
near Peterborough
Cambridgeshire, England

FOSSIL REMAINS: A single
left leg bone

CETIOSAURISCUS
LIKE WHALE LIZARD

Inhabiting the same coastal forests once roamed by the earlier Cetiosaurus, and looking a bit like it, Cetiosauriscus was another dinosaur that dined among the treetops and walked around on four pillar-like legs. If the two could have been placed side-by-side however, some differences would have been immediately obvious.

Cetiosauriscus was smaller, slimmer and more lightly built than Cetiosaurus. Its tail may also have been longer – although this is uncertain. Taken together, these features might seem to make Cetiosauriscus appear less advanced in a world where dinosaurs everywhere were getting bigger and heavier, but in fact its structure seems to have made it more successful. Later long-necked dinosaurs of a similar form tended to be more like Cetiosauriscus than Cetiosaurus.

Successful as it was, Cetiosauriscus may have been preyed upon by the large meat-eater Eustreptospondylus – fossils of which were found in roughly the same place.

The animal was named Cetiosauriscus – 'Like Cetiosaurus' – because German palaeontologist Friedrich von Huene thought it looked so similar to Cetiosaurus fossils already found that it was probably a sub-species of the same type.

More recently though, it has been determined that Cetiosauriscus lived more recently and was definitely a distinct species in its own right.

VITAL STATISTICS

NAME: Cetiosauriscus
(see-tee-oh-sore-is-kuss)

NAME MEANING: Like Cetiosaurus
(Whale Lizard)

FAMILY: Diplodocoidea

ESTIMATED SIZE: 15m long

ESTIMATED WEIGHT: 4 tonnes

DIET: Herbivore

ANATOMICAL CHARACTERISTICS: Lightly
built, possibly with very long tail

LIVED DURING: Callovian stage of the
Mid-Jurassic

WHERE FOUND: Oxford Clay
Formation, near Fletton, Peterborough,
Cambridgeshire, England

FOSSIL REMAINS: Front leg, back leg,
backbone and parts of pelvis

EUSTREPTOSPONDYLUS
WELL-CURVED VERTEBRA

Stalking the coastal forests of the islands that made up what is now central England, Eustreptospondylus was a large meat-eater. Its prey consisted of the many herbivores that browsed among the trees in the same area.

Smaller prey such as Callovosaurus may have been able to evade it through speed and stealth, while contemporary medium-sized animals such as Lexovisaurus and Sarcolestes had developed defences which would have made a hungry Eustreptospondylus think twice before taking them on.

Larger animals such as Cetiosauriscus would have been too difficult for Eustreptospondylus to handle though it would have been able to scavenge from their carcasses. As a coastal animal, Eustreptospondylus would also have been able to feast on the bodies of large ocean-going reptiles and other creatures washed up along the shoreline.

Moving about on two legs with its long, stiff tail stretched out behind for balance, Eustreptospondylus used its formidable jaws and sharp teeth as its main weapon but could also use its short arms and claws to inflict damage too.

It has also been suggested that Eustreptospondylus was a good swimmer – able to move from island to island using a combination of its powerful legs and arms. Taking to the water during the Callovian stage of the Mid-Jurassic era might have been hazardous however, with large carnivorous marine reptiles such as the 6m long Liopleurodon roaming the seas.

VITAL STATISTICS

NAME: Eustreptospondylus
(ewe-strep-toe-spon-die-luss)

NAME MEANING: Well-Curved Vertebra

FAMILY: Eustreptospondylinae

ESTIMATED SIZE: 5-7m long

ESTIMATED WEIGHT: 500kg

DIET: Carnivore

ANATOMICAL CHARACTERISTICS:
Powerful legs, short but useful arms
with claws

LIVED DURING: Callovian stage of the
Mid-Jurassic

WHERE FOUND: Oxford Clay Formation,
Webb's Pit, Summertown, Wolvercote,
Oxford, Oxfordshire, England

FOSSIL REMAINS: Partial skull, backbone,
tail, legs and feet, right arm

LEXOVISAURUS
LEXOVII LIZARD

The Lexovii Lizard shares many characteristics with its Stegosaur cousin, namely the same overall shape with a low head and heavily-armoured neck, back, shoulders and flexible and very strong tail.

Travelling on all four legs, Lexovisaurus roamed the great areas of vegetation that covered much of the south of England. These were pack animals and lived and travelled in groups.

One of the most striking features of this dinosaur is the position of large spikes – one on each shoulder – plus the more familiar tail spikes that follow the typical stegosaurian family pattern. When Lexovisaurus was first discovered in 1888 there was much confusion about exactly where the shoulder spikes came from. This was because the spikes were not attached directly to the dinosaur's skeleton; instead they were connected directly to the skin and when the animal died the spikes came away from the body making positioning them today extremely difficult.

At 6m long and weighing in at around a tonne, these were no dinosaurs to be trifled with and predators had to work out a strategy to get an individual away from the pack if they were going to feast on Lexovisaurus. stegosaurians of all sizes would often stand together and swing their large, stout, spined tails at any attacker.

The name Lexovisaurus – 'Lexovii Lizard' – is from a Celtic tribe that lived in France during Roman times. The country of the Lexovii was one of the places from which the Romans crossed the Channel to reach Britain.

While Lexovisaurus is still considered to be a distinct species by the Natural History Museum in London, it has been suggested lately that its name should be changed to Loricatosaurus, meaning 'Ancient Armoured Lizard'.

VITAL STATISTICS

NAME: Lexovisaurus
(leck-so-vee-sore-us)

NAME MEANING: Lexovii Lizard

FAMILY: Stegosauridae

ESTIMATED SIZE: 6m long

ESTIMATED WEIGHT: 1 tonne

DIET: Herbivore

ANATOMICAL CHARACTERISTICS: General
quadroped Stegosaurian shape with low
hung head, flat plates along the back
and spines that site all along the tail,
spikes sticking out from the shoulders

LIVED DURING: Callovian stage of the
Mid-Jurassic

WHERE FOUND: Tanholt village near Eye,
Cambridgeshire

FOSSIL REMAINS: Five parts of spine,
pieces of armour and limb bones

STEGOSAURUS PRISCUS
ROOFED LIZARD

Stegosaurus priscus was an average-sized dinosaur that roamed throughout the large western woodlands of Europe. It was a herbivore that lived in herds to protect its young.

It would have had atypical kite-shaped spine plates on its neck and back with smaller forelimbs than hindlegs giving it a lower position at the front. Its head was narrow and long and comparatively small in relation to the rest of its body.

Large, flat teeth were positioned towards the back of its jaw but there were no teeth at the front of its mouth because like others of the same family, Stegosaurus priscus effectively had a beak.

There were four toes on each of the hind feet and five toes on the fore feet although on the latter the inner two toes were virtually fused into a blunt hoof.

It was originally thought to have a large spine on each shoulder but later study found that these spines came from within the arrangement of spikes towards the end of the tail. This group of formidable spikes, often used as an organic and extremely powerful mace, would be swung at predators with force – the grouping of spikes being known as a Thagomizer.

VITAL STATISTICS

NAME: Stegosaurus priscus
(stegg-o-sore-us-priss-kuss)

NAME MEANING: Roofed Lizard

FAMILY: Stegosauridae

ESTIMATED SIZE: 4m long

ESTIMATED WEIGHT: 800kg

DIET: Herbivore

ANATOMICAL CHARACTERISTICS: Small
head, large rounded body, forelimbs
shorter than rear legs with large spinal
plates and spikes on the tail to the end

LIVED DURING: Callovian stage of the
Mid-Jurassic

WHERE FOUND: Fletton brick pit,
Peterborough, England

FOSSIL REMAINS: Partial
cranial specimens

SARCOLESTES
FLESH ROBBER

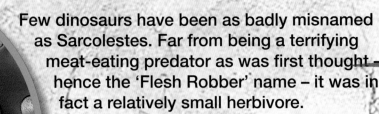

Few dinosaurs have been as badly misnamed as Sarcolestes. Far from being a terrifying meat-eating predator as was first thought – hence the 'Flesh Robber' name – it was in fact a relatively small herbivore.

That's not to suggest that Sarcolestes was an easy meal for any marauding carnivore; it was well defended with thick, bony plates of armour across its back, particularly over its vulnerable neck. Its long tail ended in a dense bone club that could be used to deter attackers.

Unlike some other members of the ankylosaur family, Sarcolestes was very low to the ground, walking on all fours, and would have fed exclusively on young plants, low branches and possibly roots. It used the beak-like front part of its mouth to snip bits off before its rows of stubby serrated teeth then ground them up.

Sarcolestes' inappropriate name was bestowed by English naturalist and geologist Richard Lydekker in 1893.

Having been reassessed, it was realised that Sarcolestes' jawbone certainly belonged to a plant-eater but it was thought to be that of a stegosaur. It wasn't until 1983 that Sarcolestes was considered to be an ankylosaur instead.

VITAL STATISTICS

NAME: Sarcolestes (sar-ko-less-teez)

NAME MEANING: Flesh Robber

FAMILY: Nodosauridae

ESTIMATED SIZE: 3m long

ESTIMATED WEIGHT: 150kg

DIET: Herbivore

ANATOMICAL CHARACTERISTICS:
Relatively small with armour plates
and tail club

LIVED DURING: Callovian stage of the
Mid-Jurassic

WHERE FOUND: Brick pit at Fletton,
near Peterborough, Cambridgeshire

FOSSIL REMAINS: Part of lower left
jawbone

METRIACANTHOSAURUS
MODERATELY-SPINED LIZARD

Metriacanthosaurus was a large, fast predator that looked similar to the Megalosaurs of the Late Jurassic period except for the bones of its spine, which gave it a slightly taller back. It hunted in packs and would prey largely on smaller slower-moving herbivores, rather than tackling the gigantic plant-eaters that lived alongside them. Being considerably smaller than other carnivores of later eras was an advantage because it allowed the dinosaur to be fast through the dense areas of woodland mainly in southern England. These dinosaurs roamed the land acting like modern-day wolves, isolating animals from their herds before tearing them to shreds.

This dinosaur was long thought to be part of the Megalosaurus family until it was discovered that the animal's vertebrae were considerably taller than others in that family. Metriacanthosaurus gets its name from those vertebrae which are taller than those in Allosaurus but lower than other high-spined dinosaurs like Acrocanthosaurus.

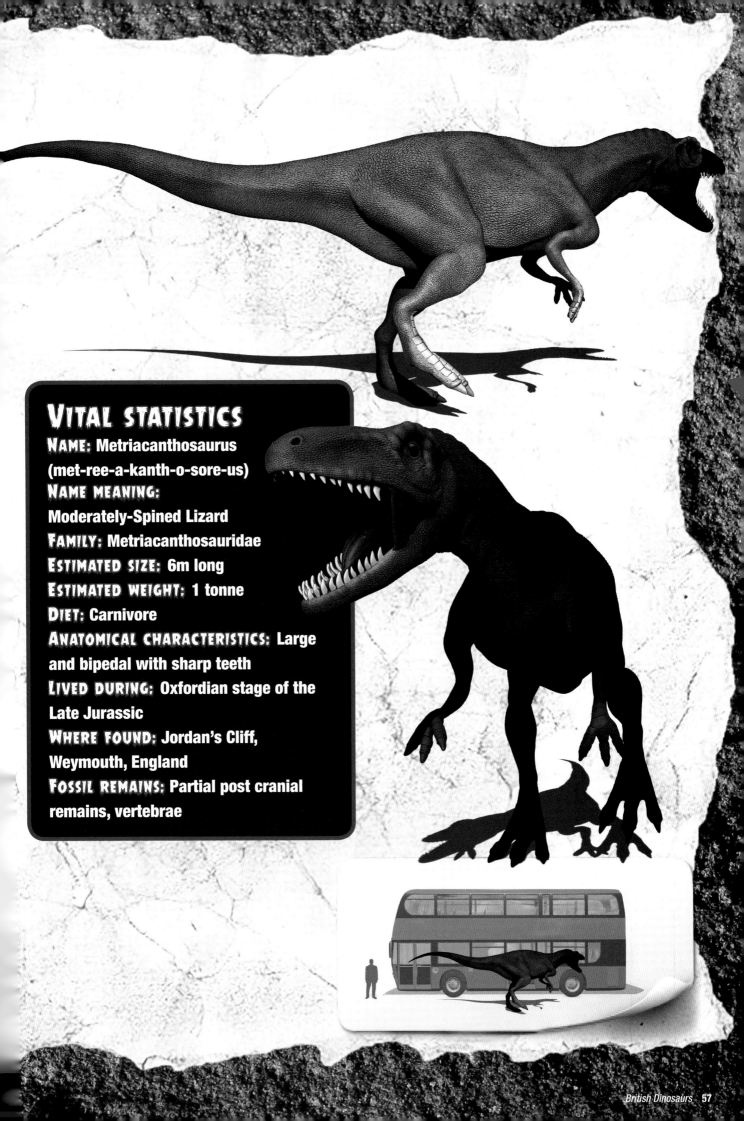

VITAL STATISTICS

NAME: Metriacanthosaurus
(met-ree-a-kanth-o-sore-us)

NAME MEANING:
Moderately-Spined Lizard

FAMILY: Metriacanthosauridae

ESTIMATED SIZE: 6m long

ESTIMATED WEIGHT: 1 tonne

DIET: Carnivore

ANATOMICAL CHARACTERISTICS: Large
and bipedal with sharp teeth

LIVED DURING: Oxfordian stage of the
Late Jurassic

WHERE FOUND: Jordan's Cliff,
Weymouth, England

FOSSIL REMAINS: Partial post cranial
remains, vertebrae

DACENTRURUS
POINTED TAIL

Woodland-dwelling Dacentrurus specialised in eating the vegetation that grew at ground level. It was one of the largest Stegosaur-type dinosaurs, with unusually long front legs compared to similar species.

It is believed to have had shoulder spikes similar to those of the African Kentrosaurus and these, along with heavy armour, would have acted as a significant deterrent to predators.

The dinosaur most likely to pose a threat to Dacentrurus was probably Juratyrant – a 5m long carnivore which stalked the same islands and woodlands as the herbivore.

Living alongside Dacentrurus were more long-necked treetop browsers such as Duriatitan and the Iguanodon-like Cumnoria.

Dacentrurus was the first Stegosaur-type dinosaur to be discovered but a combination of unfortunate circumstances meant that while the Stegosaurus went on to worldwide fame Dacentrurus was left to languish all but unknown in the shadows.

The first fossils of Dacentrurus were named in 1875 by Richard Owen but he called it Omosaurus – 'Upper Arm Lizard'. This name, it turns out, was already taken by a different species. Two years later, famous and widely publicised American palaeontologist Othniel Charles Marsh found an almost complete fossil skeleton of what he called Stegosaurus. Omosaurus was left looking for a new name, and the rest was history.

VITAL STATISTICS

NAME: Dacentrurus (dah-sen-troo-russ)

NAME MEANING: Pointed Tail

FAMILY: Stegosauridae

ESTIMATED SIZE: 6-10m long

ESTIMATED WEIGHT: 2-5 tonnes

DIET: Herbivore

ANATOMICAL CHARACTERISTICS: Sharp back plates becoming tail spikes

LIVED DURING: Kimmeridigian stage of the Late Jurassic

WHERE FOUND: Swindon Brick and Tile Company brick pit, Swindon, Wiltshire, England

FOSSIL REMAINS: Hip bones, parts of legs and parts of backbone

CUMNORIA
FROM CUMNOR

Standing on two legs, Cumnoria was able to nip at higher branches with its beak-like mouth, or on all fours it could reach down to nibble ground-level foliage.

It had no teeth at the sharp front of its mouth but did have strong ones at the sides so that it could chew up the plant material it had snipped off.

Living at the same time as Dacentrurus, it too would probably have had cause to fear predators such as Juratyrant. Lacking armour, weapons or sheer size as a defence, Cumnoria probably had to rely on speed for survival.

This adaptability in feeding and mobility in avoiding enemies probably explains why the Iguanodon-like species was so successful. While Cumnoria lived in what is now the area west of Oxford, similar animals called Camptosaurus roamed much further afield – some examples having been found in what is now North America.

When it was first discovered by Professor Joseph Prestwich, Cumoria was thought to be an Iguanodon. Then it was thought a Camptosaurus, as those found in America, but today it is regarded as a separate species in its own right – though probably similar to Camptosaurus in appearance.

VITAL STATISTICS

NAME: Cumnoria (come-nor-ee-ah)

NAME MEANING: From Cumnor

FAMILY: Ankylopollexia

ESTIMATED SIZE: 3-3.5m long

ESTIMATED WEIGHT: 55-80kg

DIET: Herbivore

ANATOMICAL CHARACTERISTICS: Beak-like mouth front, 'hands' for gripping food

LIVED DURING: Kimmeridgian stage of the Late Jurassic

WHERE FOUND: Chawley Brick Pits, Cumnor Hurst, Oxfordshire

FOSSIL REMAINS: Large parts of skeleton including tail, backbone, legs, arms and parts of skull

Before the Great Extinction
The Cretaceous Period
[142 to 65 million years ago]

At the end of the Jurassic going into the start of the Cretaceous the sea had receded enough to connect the southernmost of the British islands with the rest of Europe.

Much of the rest of the country, however, remained a series of islands and the connection eventually disappeared to make England and Scotland unconnected islands once again as sea levels rose.

Early Cretaceous Britain was populated by even more species than ever before. The first bees began to appear during the Cretaceous, along with other familiar insects such as ladybirds, cockroaches and aphids although it is unknown precisely when these species arrived in what is now Britain.

Whatever species of insects were living in Britain during this period, they would have encountered the first flowering plants – giving some real colour to the landscape for the first time. The earliest known British fossil flower, a tiny specimen, was found in southern England and dates from the Early Cretaceous. During this time, the area was a humid and sub-tropical floodplain which probably alternated between wet rainy seasons and very dry summers. Evidence exists of fires sweeping across the area but there was still no grass on the ground.

The tall forests of earlier Britain had probably been replaced by low-lying vegetation and large areas of flowing water inhabited by a variety of fish. Out at sea, ichthyosaurs were steadily dying out – while plesiosaurs, though still successful, were beginning to face competition from what would eventually become the most powerful underwater predators of the Late Cretaceous – Mosasaurs.

In the skies, the tailed pterosaurs had been superseded by much larger tailless species such as Ornithocheirus – 'Bird Hand' – which could grow up to 5m in wingspan.

On land, Britain was ruled by the dinosaurs. Some of the largest known British species appeared during the Cretaceous, such as the fish-eating carnivore

A map of the British Isles showing, in green, areas of rock that have survived from the Cretaceous period.

The islands of Britain were beginning to more closely resemble their modern day counterparts during the Cretaceous period.

Baryonyx and huge sauropods such as Luticosaurus. However, there was also a profusion of smaller species, both meat-eaters and herbivores. The former had to compete for their food with crocodiles, at least eight species of which made southern England their home at this time.

Perhaps the most successful dinosaurs of all during the Early Cretaceous were the ornithopods – the most famous being Iguanodon. These medium-size animals were big enough to deter predators, could reach up high or down low for their food and most importantly of all were able to chew it.

This meant they digestion was greatly speeded up and they didn't need huge guts like long-necked sauropods to break down their food over several days.

A strange offshoot of dinosaurs also appeared during the Cretaceous. These slender lightweight feathered creatures had long beak-like mouths for more easily catching small prey such as fish or insects. Some were able to swim and dive while others were actually able to fly like pterosaurs. These were the first birds.

During the Late Cretaceous period, up to the extinction of the dinosaurs 65 million years ago, some of the most famous dinosaurs roamed the earth – Tyrannosaurus Rex, Triceratops and Velociraptor for example – and Britain may well have been inhabited by something similar, but we just can't say for certain because Britain has very few Late Cretaceous fossil-bearing rocks. These have mostly already eroded away, leaving us with only the tantalising possibility of what might have been, represented by fossils from that period found elsewhere in the world.

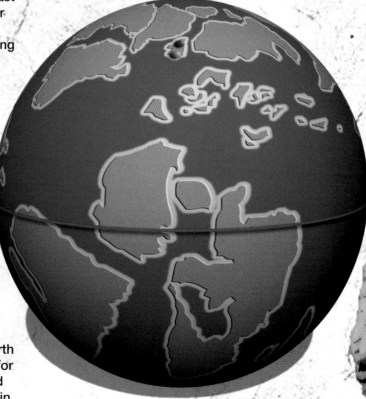

The world as it appeared during the Late Cretaceous period with Scotland in blue and England in red.

ECHINODON
SPINY TOOTH

One of the smallest dinosaurs ever discovered, Echinodon lived in a world where there were plenty of places to hide from large predators. Its small size would have enabled it to disappear among the dense foliage that covered what is now southern England.

As a plant-eater, it would have had to eat tender shoots or possibly roots since its narrow blade-like teeth were not adapted for grinding and chewing. An unusual feature of Echinodon was its matching sets of canine-like teeth. These have given rise to speculation that it may in fact have been an omnivore – supplementing its diet with small mammals or lizards.

Little though it may have been, it is unlikely that Echinodon would have escaped the notice of predatory hunters such as Nuthetes.

At 55-60cm in length, Echinodon may have been the smallest dinosaur of its age but it was not the smallest of all time. A whole host of tiny dinosaurs, most of them feathered, unlike Echinodon, have been discovered. These include the 25cm long Epidexipteryx, 30cm Eosinopteryx and 30-39cm Parvicursor remotus.

Named in 1861 by Richard Owen, Echinodon was among the earlier dinosaur discoveries, although it was initially mistaken for a lizard due to its size.

VITAL STATISTICS

NAME: Echinodon (eh-ky-no-don)

NAME MEANING: Spiny Tooth

FAMILY: Heterodontosauridae

ESTIMATED SIZE: 55-60cm long

ESTIMATED WEIGHT: 2-4.5kg

DIET: Herbivore

ANATOMICAL CHARACTERISTICS: Very small but otherwise too little known

LIVED DURING: Berriasian stage of the Early Cretaceous

WHERE FOUND: Durlston Bay, Isle of Purbeck, Dorset, England

FOSSIL REMAINS: Jawbone fragments, parts of skull

Nuthetes Destructor
Admonisher

Nuthetes may have been a small dinosaur, about 2.3m long and weighing around 140kg, but it was a fast and vicious pack hunter with sharp, serrated teeth.

At least partially feathered, with an elaborate tail and small, stubby wings that were more of a mid-way between lizard and bird, Nuthetes stalked its prey through the undergrowth.

The precise purpose of its feathers is unknown, but they may have been either for insulation or display and courtship.

Parts of teeth fragments found on some fossil reptilian and mammalian remains of the time suggest that Admonisher (or Monitor) was adept at picking off other animals near the water's edge of small rivers or could actually take animals from shallow water if needed.

Nuthetes destructor was identified as a carnivore from teeth and jaw fragments found in Swanage. Indeed, the name comes from the similarity between the adaptations to the teeth for piercing, cutting and lacerating the prey animal flesh and those of a modern large Bengal monitor lizard.

It was identified and named by Richard Owen in 1854.

VITAL STATISTICS

NAME: Nuthetes (noo-thay-tees)

NAME MEANING: Admonisher (or Monitor)

FAMILY: Dromaeosauridae

ESTIMATED SIZE: 2.3m long

ESTIMATED WEIGHT: 140kg

DIET: Carnivore

ANATOMICAL CHARACTERISTICS: Small bipedal predator with sharp teeth, partially feathered

LIVED DURING: Middle Berriasian stage of the Early Cretaceous

WHERE FOUND: Durlston Bay, Isle of Purbeck, Dorset, England

FOSSIL REMAINS: Teeth and jaw fragments

BECKLESPINAX
BECKLE'S SPINE

Tall meat-eating Becklespinax stalked its prey on its hind legs but probably also had powerful forelimbs. It lived at the same time as Craterosaurus and Hylaeosaurus and may have hunted them. Then again, the sail on Becklespinax's back raises the possibility that it might have been adapted to spend some of its life in the water looking for fish.

The most famous meat eater with a sail on its back, Spinosaurus, is known to have eaten fish and although it lived later than Becklespinax it was not that far away in what is now North Africa. Another species closer to Becklespinax in both time and appearance, Concavenator, also had a sail on its back and what appear to be short, stiff feather-like filaments on its arms. It is possible that Becklespinax had these 'feathers' as well.

There are too few Becklespinax pieces – just three vertebrae with spines along their tops – to be certain of exactly how the rest of the dinosaur looked however. The fossil remains that we have were discovered by Samuel Husband Beckles in the 1850s and for some time it was thought that they belonged to Megalosaurus – leading to several depictions of that dinosaur with a hump on its back. It eventually received its current name in 1991.

VITAL STATISTICS

NAME: Becklespinax (beck-ull-spy-nax)

NAME MEANING: Beckle's Spine

FAMILY: Carnosauria

ESTIMATED SIZE: 5m long

ESTIMATED WEIGHT: 1 tonne

DIET: Carnivore

ANATOMICAL CHARACTERISTICS: Hump or sail on its back

LIVED DURING: Berriasian-Valanginian stage of the Early Cretaceous

WHERE FOUND: A quarry near Battle, East Sussex, England

FOSSIL REMAINS: Part of backbone

CRATEROSAURUS
BOWL LIZARD

Stegosaur-type plant-eaters were flourishing in what is now England during the time of Craterosaurus, alongside long-necked, long-tailed treetop munchers like Pelorosaurus and other armoured low browsers such as Hylaeosaurus.

Probably sporting rows of both plates and spines along its back, Craterosaurus had some protection against marauding big predators such as Valdoraptor as it moved through its woodland home eating low-growing cycads and other plants.

At only 4m long, Craterosaurus was on the small-side for a Stegosaur and more lightly built than other members of the same family. This may have given it an extra turn of speed when it came to escaping from meat-eaters.

Known only from a single piece of backbone, Craterosaurus was named by palaeontologist Harry Seeley in 1874. He thought he had found part of its skull and it wasn't until 1912, three years after his death, that Transylvanian-born aristocrat Baron Franz Nopcsa worked out that it was actually a broken piece of vertebra.

It has since been suggested that the lone Craterosaurus specimen discovered, small as it is, might actually be part of a young Regnosaurus. Not enough material remains to be sure though.

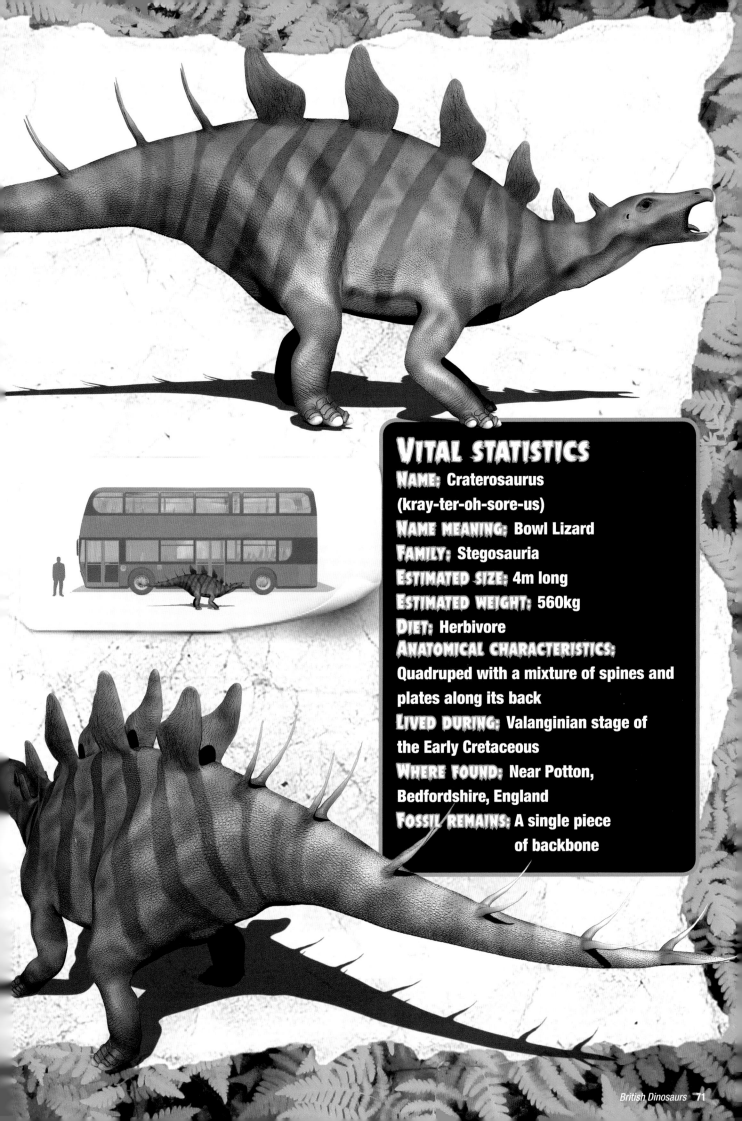

VITAL STATISTICS

NAME: Craterosaurus
(kray-ter-oh-sore-us)

NAME MEANING: Bowl Lizard

FAMILY: Stegosauria

ESTIMATED SIZE: 4m long

ESTIMATED WEIGHT: 560kg

DIET: Herbivore

ANATOMICAL CHARACTERISTICS:
Quadruped with a mixture of spines and plates along its back

LIVED DURING: Valanginian stage of the Early Cretaceous

WHERE FOUND: Near Potton, Bedfordshire, England

FOSSIL REMAINS: A single piece of backbone

HYLAEOSAURUS
WOODLAND LIZARD

As its name suggests, Hylaeosaurus was a woodland creature. It fed on the wealth of plant life that was growing at ground level in southern England during the Early Cretaceous period.

The sort of problems it faced from predators are illustrated by the ridged armour plates that grew on its back. Unlike stegosaurs of the period, however, it seems to have had no offensive weaponry in the form of spikes or a tail club.

In addition, rather than waddling along with its belly close to the ground, Hylaeosaurus had reasonably long legs which would have enabled it to move away from danger at a brisk pace. The danger – in the form of meat-eaters like Valdosaurus – might well have been able to catch up to it but it's always more difficult to bring down a moving target.

One of the earliest dinosaurs to be found and identified in England, Hylaeosaurus was named by Gideon Mantell in 1833. It was one of the trio of animals, alongside Iguanodon and Megalosaurus, first described as 'dinosaurs' by Richard Owen in 1842. It appears to be very similar to the later Polacanthus, although there are enough differences to identify Hylaeosaurus as a distinct species.

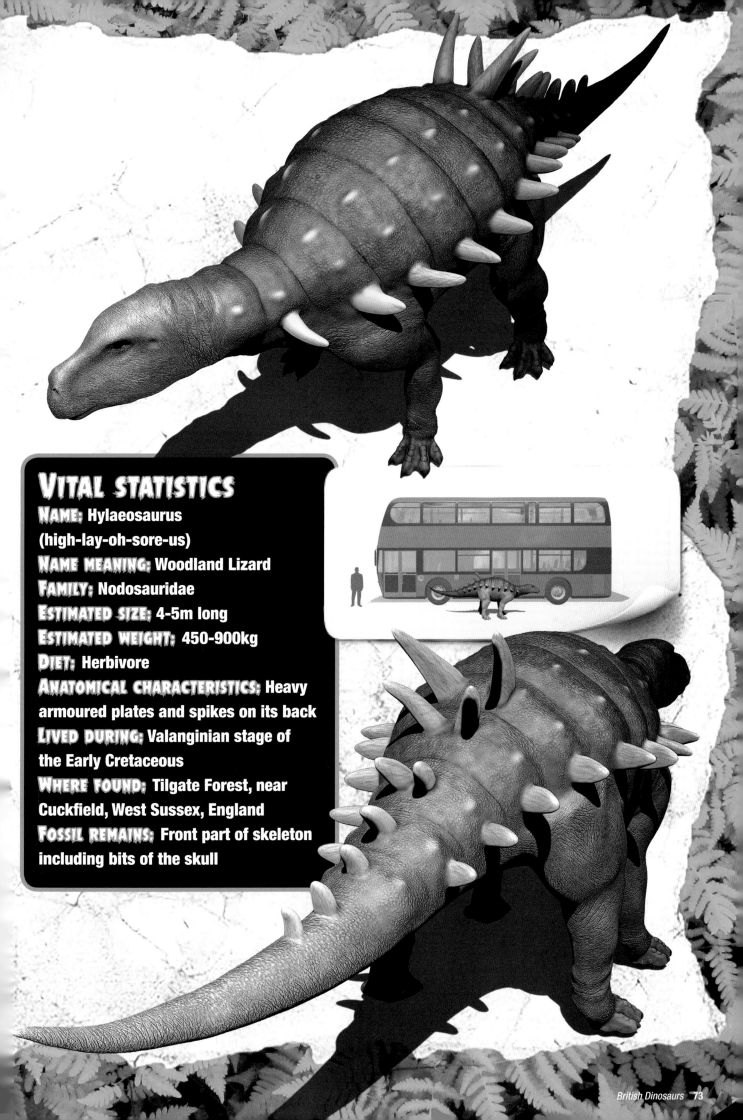

VITAL STATISTICS

NAME: Hylaeosaurus
(high-lay-oh-sore-us)

NAME MEANING: Woodland Lizard

FAMILY: Nodosauridae

ESTIMATED SIZE: 4-5m long

ESTIMATED WEIGHT: 450-900kg

DIET: Herbivore

ANATOMICAL CHARACTERISTICS: Heavy armoured plates and spikes on its back

LIVED DURING: Valanginian stage of the Early Cretaceous

WHERE FOUND: Tilgate Forest, near Cuckfield, West Sussex, England

FOSSIL REMAINS: Front part of skeleton including bits of the skull

PELOROSAURUS
MONSTROUS LIZARD

This was a true giant of its age – a huge dinosaur the very size of which resulted in its name 'Monstrous Lizard'. As large as it was, however, Pelorosaurus still had armour and tough skin to help repel attacks from carnivores.

This was likely to be of most benefit to juvenile animals growing up in the herd. Certainly, once it got to full adult size, a defensive Pelorosaurus would have been a formidable foe for any attacker. It is plausible that this dinosaur was a kind of titanosaur, given the hexagonal 'osteoderms' visible on its side that have been left in fossilised skin impressions. The armoured nodules along the animal's back, combined with the unique family skin pattern would certainly leave this dinosaur vying to be included in the titanosaur category.

Striding through the woodlands of Cretaceous Britain, Pelorosaurus was a tree-top grazer, munching its way through tons of vegetation that was out of reach to other dinosaurs. Its extremely long neck, comparatively small head and long front legs served a purpose when it came to getting at tasty young branches at the tops of trees and between the boughs to juicy new shoots.

Fossils of Pelorosaurus were among the first sauropod remains to be found and classified – by Gideon Mantell in 1850.

VITAL STATISTICS

NAME: Callovosaurus
(cal-oh-voh-sore-us)

NAME MEANING: Monstrous Lizard

FAMILY: Brachiosauridae,
Macronaria (sauropods with
nasal crest)

ESTIMATED SIZE: 24m long

ESTIMATED WEIGHT: 15 tonnes

DIET: Herbivore

ANATOMICAL CHARACTERISTICS:
Long-necked, long-tailed with oblong
body, long front legs and shorter back
ones, with hexagonal-patterned skin
and armoured nodules on its back

LIVED DURING: Valanginian stage of
the Early Cretaceous

WHERE FOUND: Sandown, Cuckfield,
West Sussex, England

FOSSIL REMAINS: Neck and back
vertebrae, arm bone, hip bones, hind
leg bones, skin impressions

PLEUROCOELUS
HOLLOW-SIDED

A herbivore that would have grazed on the tough vegetation of the woodlands that typified the area of the time, Pleurocoelus was a big animal.

Its long neck, balanced by an equally long and robust tail, allowed Pleurocoelus to reach up into the higher branches of trees looking for prime leaves and shoots. Moving in herds, they would have worked their way through dense woody areas with the larger animals providing a safe environment for the smaller dinosaurs in the family group.

In terms of British dinosaur identification, Pleurocoelus is something of a puzzle. Only one tooth has yet been found in the UK but from the shape and type of this rare find we know that it fell into a group of animals known as titanosauriformes.

The tooth has laterally placed labial grooves which helps identify it as belonging to a close relative of another British giant – the chronologically more modern Oplosaurus. Some experts believe that Oplosaurus may turn out to be an adult version of Pleurocoelus should more fossils be uncovered.

VITAL STATISTICS

NAME: Pleurocoelus (ploo-row-seel-uss)

NAME MEANING: Hollow-Sided

FAMILY: Brachiosauridae

ESTIMATED SIZE: 10m long

ESTIMATED WEIGHT: 20 tonnes

DIET: Herbivore

ANATOMICAL CHARACTERISTICS: Large with long neck, long tail, stout legs and rounded oblong body, mid-sized head with flat, plant-stripping teeth

LIVED DURING: Valanginian stage of the Early Cretaceous

WHERE FOUND: Isle of Wight, England

FOSSIL REMAINS: One tooth

REGNOSAURUS
SUSSEX LIZARD

This dinosaur is a member of the stegosaur family and shared many characteristics with its larger cousin. A herbivore, Regnosaurus roamed throughout the woodlands of what would later become Britain grazing on the rough vegetation and avoiding the more proficient predators of the time.

With shorter front legs and longer hind legs, coupled with the typical arched back under two rows of external, upright plates and spikes along its tail Regnosaurus looked very much like other stegosaurs. One difference though was its neck, which was a little longer than that of other stegosaurs. Regnosaurus had a broad skull and flat teeth housed in the front of its mouth and would have been limited to vegetation in the lower sweep of plant life – because the armour around its shoulders and the back of its neck would have seriously limited its range of movement upward.

This was a social dinosaur and would have lived in herds, moving as a group from one area of food and shelter to another as needed.

When fossils of Regnosaurus were first discovered in 1839, Gideon Mantell thought they belonged to Iguanodon, but in 1848 he changed his mind and recognised the dinosaur as a distinct species.

VITAL STATISTICS

NAME: Regnosaurus (reg-no-sore-us)

NAME MEANING: Sussex Lizard

FAMILY: Huayangosauridae

ESTIMATED SIZE: 3-4m long

ESTIMATED WEIGHT: 500kg to 1 tonne

DIET: Herbivore

ANATOMICAL CHARACTERISTICS:
Mid-sized stegosaur-type dinosaur with small head and long tail, characterised by large plates on the spine and spikes along the length of the tail

LIVED DURING: Berriasian stage of the Early Cretaceous

WHERE FOUND: Cuckfield, Sussex, England

FOSSIL REMAINS: Pubis, spikes, lower jaw fragment, teeth

Valdoraptor
Wealden Plunderer

At 6m long, Valdoraptor was a sizeable meat-eating dinosaur and one that would have been an eager predator. Standing on its powerful rear legs, Valdoraptor held its head up on a long, muscular neck.

It would have had powerful jaws in a long skull and may have ambushed its prey, rather than chasing after it. Although Valdoraptor's tail wasn't particularly thick it would have been long and used as a counterbalance when the animal ran.

Long forelimbs might have been effective for pushing vegetation out the way, grabbing hold of a potential meal or even as a secondary weapon to Valdoraptor's jaws. This dinosaur may have had an element of feathering, though how much is unclear, and possibly a crest, dorsal sail or other features.

Valdoraptor, although identified purely from a few incomplete bones from one of its feet, was probably very close to the top of the food chain in its time.

Its fossil remains were first discovered in the 1850s, when they were thought to belong to Hylaeosaurus. Next they were thought to be part of Megalosaurus and it wasn't until 1991 that Valdoraptor was finally accepted as a species of carnivore in its own right.

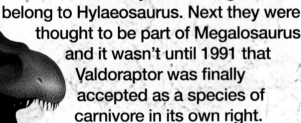

Vital statistics

Name: Valdoraptor (val-doe-rap-tor)

Name meaning: Wealden Plunderer

Family: Tetanurae

Estimated size: 6m long

Estimated weight: 2 tonnes

Diet: Carnivore

Anatomical characteristics: Large and bipedal with sharp teeth

Lived during: Valanginian stage of the Early Cretaceous

Where found: Cuckfield, West Sussex, England

Fossil remains: Three bones from the right foot

BARYONYX
HEAVY CLAW

Living alongside several species of crocodile, Baryonyx competed with them for the same sort of food. Like other Spinosaurs, it was semi-aquatic, spending some of is time on land and the rest hunting for fish in shallow pools and rivers.

Its long snout made it an effective fisher and if necessary the large, curved claw on its thumb could be used to deliver a killing blow to a struggling victim.

Baryonyx's fish diet was confirmed when fragments of bones and scales belonging to a prehistoric fish, Scheenstia, were found fossilised within the stomach of a specimen discovered in Surrey.

Remains of an Iguanodon-type dinosaur, probably Mantellisaurus, were also found, suggesting that Baryonyx was not above scavenging from the carcasses of larger animals on dry land for additional food.

Measuring some 9m long and able to walk around quickly on its large powerful hind legs, Baryonyx was probably larger than any of its competitors and existed at the top of its food chain. It remains one of the largest meat-eating dinosaurs ever found in Europe, larger even than the 8m Neovenator which existed in England during the same time period.

The best known fossil of Baryonyx was found in January 1983 by fossil collector William Walker and named Baryonyx walkeri in his honour three years later.

VITAL STATISTICS

NAME: Baryonyx (Barry-on-icks)

NAME MEANING: Heavy Claw

FAMILY: Spinosauridae

ESTIMATED SIZE: 9-10m long

ESTIMATED WEIGHT: 1.7-2.7 tonnes

DIET: Carnivore

ANATOMICAL CHARACTERISTICS: Long snout, single large thumb claw on each hand

LIVED DURING: Early Barremian stage of the Early Cretaceous

WHERE FOUND: Smokejacks Brickworks, Ockley, near Dorking, Surrey, England. Fragments also found on the Isle of Wight, in Spain and in Portugal

FOSSIL REMAINS: Large part of skeleton including legs, ribs, hips, backbone and parts of skull

ARISTOSUCHUS PUSILLUS
BRAVEST LITTLE CROCODILE

Built for speed, Aristosuchus pusillus was a small, light and agile hunter with a mouth full of sharp little teeth in a long delicate skull. It had long hind legs and a very long tail which it held out behind itself for balance as it ran.

With a diet of insects, small mammals and lizards, Aristosuchus had to be quick if it wanted a meal. Compared to earlier meat-eaters, it was highly specialised and existed at a time when many other types of dinosaur were getting larger – perhaps preventing them from taking advantage of small prey.

There were plenty of large predators around at the same time as Aristosuchus, such as Eotyrannus, Neovenator and Thecocoelurus but it is unlikely that it had much to fear from them if it kept its wits about it, being too small and fast for them to catch.

It may, however, have faced competition for food from the eight species of crocodiles that existed in the same place and at the same time – not to mention other predators of the same size such as Calamosaurus, Calamospondylus and Ornithodesmus.

Named by Richard Owen in 1876, it has been suggested that Aristosuchus looked something like the earlier and somewhat smaller Compsognathus found in Germany and France.

VITAL STATISTICS

NAME: Aristosuchus pusillus (Ah-riss-toe-soo-kuss puss-ill-us)

NAME MEANING: Bravest Little Crocodile

FAMILY: Compsognathidae

ESTIMATED SIZE: 2m long

ESTIMATED WEIGHT: 7kg

DIET: Carnivore

ANATOMICAL CHARACTERISTICS: Lightly built with long legs and tail

LIVED DURING: Barremian stage of the Early Cretaceous

WHERE FOUND: Brighstone Bay, Isle of Wight, England

FOSSIL REMAINS: Part of hip and backbone

CALAMOSAURUS
REED LIZARD

There was fierce competition for small prey during the time of Calamosaurus but while many of the larger carnivores lurked in or beside rivers and ponds, this fast agile meat-eater stalked wooded areas in search of smaller dinosaurs, lizards and insects.

Walking on two legs and coated with a mat of feather-like filaments, Calamosaurus had long hands for gripping its food and may also have had vicious claws on its feet to help it deliver a fatal blow to a fleeing herbivore.

Likely prey would have included Hypsilophodon – which was also adapted to move quickly.

This seems to have been an evolutionary arms race where both predators and prey grew ever lighter and faster to achieve greater and greater speeds – for pursuit and for escape respectively.

Other plant-eaters of the period had either grown too large for the likes of Calamosaurus to tackle, such as Iguanodon, or were small but too well defended, as was the case with Polacanthus.

Calamosaurus was only known from two pieces of backbone found in the late 1880s on the Isle of Wight until April 2015 when a third piece of backbone – a neck vertebra – was discovered by fossil hunter Dave Badman near Chilton Chine, also on the Isle of Wight.

VITAL STATISTICS

NAME: Calamosaurus (Cal-am-oh-sore-us)

NAME MEANING: Reed Lizard

FAMILY: Compsognathidae

ESTIMATED SIZE: 2-3m long

ESTIMATED WEIGHT: 70kg

DIET: Carnivore

ANATOMICAL CHARACTERISTICS: Small head, feathers, long hands

LIVED DURING: Barremian stage of the Early Cretaceous

WHERE FOUND: Isle of Wight, England

FOSSIL REMAINS: Three pieces of backbone

Calamospondylus
Reed Vertebra

Hunter Calamospondylus was a small dinosaur in a big dinosaur's world. It lived alongside some of the great giants of the early Cretaceous period such as Iguanodon and Baryonyx but its attention was probably focused on very small prey, such as insects.

It most likely spent its time living on the forest floor and may have had feathers for use during mating displays. Like the other small predators living at the same time, Calamospondylus probably faced stiff competition for food from the many other meat-eaters living in the same area.

It has been suggested that Calamospondylus could climb and spent much of its time in trees, or that it could leap around like a grasshopper but it is not possible to say exactly how it lived because the only fossil found has been lost and no drawings of it survive for modern researchers to study.

It was named in 1866 by amateur palaeontologist the Reverend William D Fox. A collection of some 500 of his specimens was acquired by the Natural History Museum after his death in 1881 but it is not known whether Calamospondylus was among them, nor what the eventual fate of the fossil was.

VITAL STATISTICS

NAME: Calamospondylus
(Cal-am-oh-spon-dee-luss)

NAME MEANING: Reed Vertebra

FAMILY: Coeluridae

ESTIMATED SIZE: 2m long

ESTIMATED WEIGHT: 20kg

DIET: Carnivore

ANATOMICAL CHARACTERISTICS: Long
fingers and claws, feathers

LIVED DURING: Barremian stage of the
Early Cretaceous

WHERE FOUND: Isle of Wight, England

FOSSIL REMAINS: A backbone section
with ribs and bits of hip – now lost

CHONDROSTEOSAURUS
CARTILAGE AND BONE LIZARD

Just like other titanosaurs, Chondrosteosaurus was a high browser with a long neck to reach the tops of trees with its small head and munch its way through dense forests using rows of spoon-shaped teeth.

Unfortunately, the area around the Isle of Wight during the Early Cretaceous was not heavily forested – certainly not enough to keep a hungry Chondrosteosaurus and its herd happy. It seems likely, therefore, that Chondrosteosaurus was a visitor to England, rather than a permanent resident.

Arriving here from feeding grounds further to the east, it would have found a host of relatively small predators vying with one another for fish from rivers and streams, and preying upon the herds of much smaller dinosaurs and other creatures that lived among the low-level vegetation.

Without a constant supply of its usual tough and fibrous food to keep it going, the Chondrosteosaurus found on the Isle of Wight may have succumbed to the fate of several other titanosaurs found in similar circumstances, such as Luticosaurus, and died of starvation.

Two pieces of Chondrosteosaurus's backbone were found during the early 1870s at an unknown location on the island's south coast before being formally named by Richard Owen in 1876.

VITAL STATISTICS

NAME: Chondrosteosaurus
(con-dross-tee-oh-sore-us)

NAME MEANING: Cartilage and Bone Lizard

FAMILY: Titanosauriform

ESTIMATED SIZE: 18m long

ESTIMATED WEIGHT: 5 tonnes

DIET: Herbivore

ANATOMICAL CHARACTERISTICS: Long
neck, long tail, large legs
and body

LIVED DURING: Barremian stage of the
Early Cretaceous

WHERE FOUND: South coast of the
Isle of Wight

FOSSIL REMAINS: Parts of backbone

EOTYRANNUS
DAWN TYRANT

Chasing after rapidly fleeing smaller dinosaurs such as Hypsilophodon was Eotyrannus's main occupation. A large meat-eater – perhaps the largest of its day – it overmatched the likes of Calamosaurus through sheer size and ferocity.

The environment in which it lived teemed with smaller animals and it may well have eaten both herbivores and little carnivores if it could get its large jaws on them.

As its name suggests, Eotyrannus was an early member of the family of dinosaurs that would eventually produce Tyrannosaurus Rex. It had a similarly shaped robust skull and long hind legs but it differed in having much longer arms in comparison.

These ended in sharp claws and Eotyrannus may have used them as a secondary weapon to help those big teeth finish the job of subduing its prey.

The only example of Eotyrannus known was found in 1997 and formally named in 2001, making it one of Britain's most recent dinosaur discoveries. This fossil skeleton was actually that of a juvenile which measured 4m from its nose to the tip of its tail – so the fully grown adult may have been a lot bigger.

Exactly how much bigger is unknown but it is unlikely to have reached the 12-13m of Tyrannosaurus Rex.

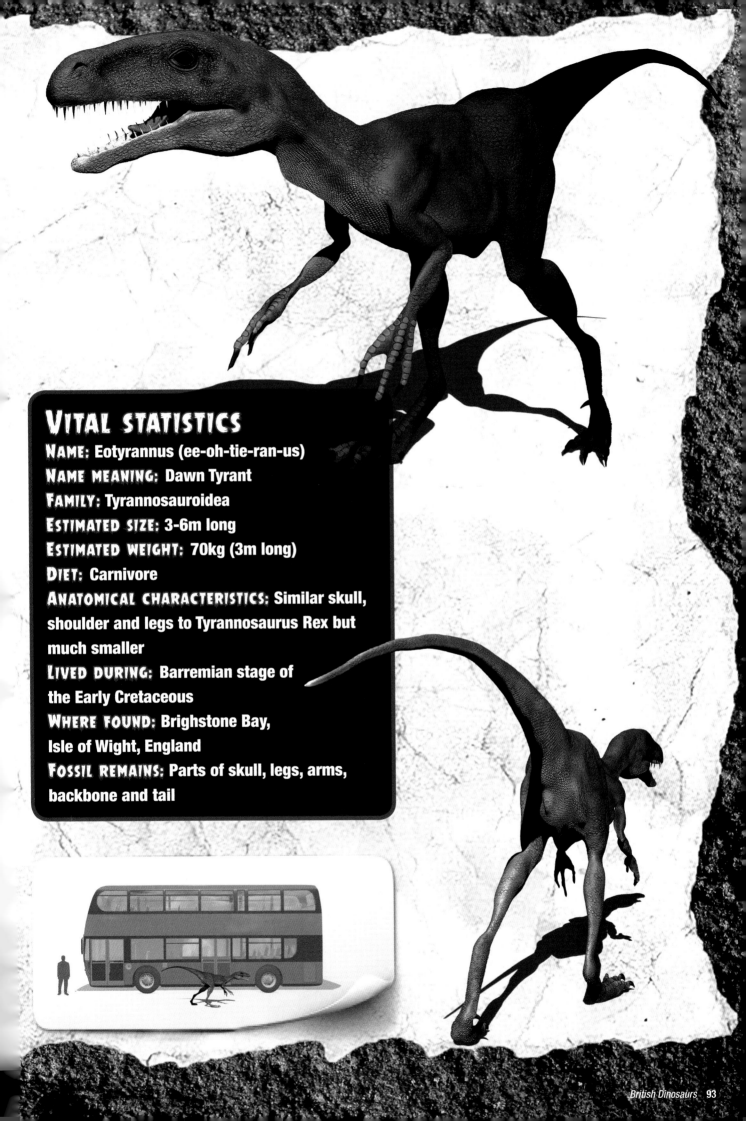

VITAL STATISTICS

NAME: Eotyrannus (ee-oh-tie-ran-us)

NAME MEANING: Dawn Tyrant

FAMILY: Tyrannosauroidea

ESTIMATED SIZE: 3-6m long

ESTIMATED WEIGHT: 70kg (3m long)

DIET: Carnivore

ANATOMICAL CHARACTERISTICS: Similar skull, shoulder and legs to Tyrannosaurus Rex but much smaller

LIVED DURING: Barremian stage of the Early Cretaceous

WHERE FOUND: Brighstone Bay, Isle of Wight, England

FOSSIL REMAINS: Parts of skull, legs, arms, backbone and tail

EUCAMEROTUS
WELL CHAMBERED

Another treetop-eating dinosaur far from home, Eucamerotus was smaller than many of its contemporaries at just 15m long and weighing in the region of 10 tonnes.

Nevertheless, just like the larger animals such as Luticosaurus and Chondrosteosaurus, it struggled to find enough to eat when it reached England during a trek westwards. Whether it starved to death is unknown but it certainly would not have found food plentiful.

It is possible that herds of Eucamerotus travelled alongside herds of the other large plant-eaters in search of better browsing.

Unfortunately, its smaller size may have made it more of a target for large predators such as Eotyrannus and it was unable to make a quick getaway like the meat-eater's usual prey.

The fossils that were named as belonging to Eucamerotus were discovered on the Isle of Wight in the early 1870s by John Whitaker Hulke, a fossil collector whose day job was working as a surgeon, and formally named by him in 1872.

The remains of another sauropod similar to Eucamerotus were found at Barnes High, also on the Isle of Wight, in 1992. These are known as the 'Barnes High sauropod' and have yet to be firmly identified.

VITAL STATISTICS

NAME: Eucamerotus (you-cam-er-oh-tus)

NAME MEANING: Well Chambered

FAMILY: Brachiosauridae

ESTIMATED SIZE: 15m long

ESTIMATED WEIGHT: 10 tonnes

DIET: Herbivore

ANATOMICAL CHARACTERISTICS: Long neck, long tail, large body and legs

LIVED DURING: Barremian stage of the Early Cretaceous

WHERE FOUND: Brook Bay, Isle of Wight, England

FOSSIL REMAINS: Parts of backbone and leg

HYPSILOPHODON
HIGH-RIDGED TOOTH

At just 2m long or smaller, large herds of Hypsilophodon grazed on tender shoots and roots across the area that is now the Isle of Wight.

It had a horny beak at the front of its mouth for cutting through vegetation and broader, stronger teeth along the sides for chewing up its food.

With five fingers on each hand, including an opposable fifth digit, it was able to grab its food and guide it into its mouth. It had a short, high skull and there is evidence that, unlike many other species of dinosaur, there would have been a visual difference between male and female animals in terms of size and other features.

As well as the safety in numbers it gained from living in large herds, Hypsilophodon was also built for speed so it could escape from the relatively high number of predators with which it shared its habitat.

Its body was light and it had a long tail which it would have stretched out stiffly for balance when running. It would have been remarkably fast for its size when on the move.

It was originally thought that Hypsilophodon lived in trees but it has since been demonstrated that it moved around rapidly on the ground instead.

The only known fossils all come from the same metre-thick 1200m long strip of mudstone on the Isle of Wight known as the Hypsilophdon bed. Dozens of individuals died and were preserved there – though not all at once. It has been suggested that it was an area of quicksand where animals, particularly smaller ones, became trapped.

VITAL STATISTICS

NAME: Hypsilophodon
(Hip-see-loaf-oh-don)

NAME MEANING: High-Ridged Tooth

FAMILY: Hypsilophodontidae

ESTIMATED SIZE: 1.9m long

ESTIMATED WEIGHT: 20kg

DIET: Herbivore

ANATOMICAL CHARACTERISTICS: Small and lightly built

LIVED DURING: Barremian stage of the Early Cretaceous

WHERE FOUND: Cowleaze Chine, Isle of Wight, England

FOSSIL REMAINS: Several complete or nearly complete skeletons

IGUANODON

IGUANA TOOTH

One of the most common dinosaurs of its age, Iguanodon was a powerful hulking herbivorous animal that was equally at home browsing on trees or low-growing ferns. This ability to reach plants across a wide height range probably contributed to its success.

Iguanodon's head was long and narrow, ending in a beak that could cut through tough vegetation but with teeth at the sides that were well adapted for chewing.

It could walk on its hind legs alone or all fours, preferring the latter, and it had large conical spikes on its thumbs. The exact function of these is uncertain but they may have been useful as a weapon – Iguanodon certainly having arms long enough to reach out and use them to cause some damage – or for foraging.

It has even been suggested that the spikes were used for fighting other Iguanodons, perhaps in mating contests.

First discovered in quarries in the area around Cuckfield, West Sussex, in 1822, Iguanodon was given its name by Gideon Mantell in 1825. This made it only the second dinosaur to be named, after Megalosaurus.

Only fragments of Iguanodon were found in England and it took some time for palaeontologists to even establish what it looked like. Famously, early reconstructions imagined that its thumb spike was actually a horn that protruded from the end of its nose. Then, in 1878, 38 Iguanodon skeletons were found lying together in a Belgian coal mine some 320m below ground. This allowed complete skeletons to be assembled.

Today, Iguanodon fossils are still the most commonly found on the Isle of Wight – an area famed for its fossils.

VITAL STATISTICS

NAME: Iguanodon (Ig-wah-no-don)

NAME MEANING: Iguana Tooth

FAMILY: Iguanodontidae

ESTIMATED SIZE: 10-12m long

ESTIMATED WEIGHT: 3.5 tonnes

DIET: Herbivore

ANATOMICAL CHARACTERISTICS: Heavily built with large thumb spike on each hand

LIVED DURING: Barremian stage of the Early Cretaceous

WHERE FOUND: Several locations along the south coast of the Isle of Wight, England, and in Belgium

FOSSIL REMAINS: Parts of skeleton including backbone, legs, hips and teeth

LUTICOSAURUS
JUTE LIZARD

Luticosaurus was a giant of a dinosaur and this may well explain why it was found in England. A great roamer around the vast forested areas of western Europe, Luticosaurus would travel in herds looking for its next meal.

At 20m long and weighing in at 26 tonnes, this was a true colossus. Indeed, Luticosarus is a member of the Titanosaur family and deserves to be noted for its sheer bulk.

Although just three incomplete pieces of backbone have been found it is widely believed that, because of its size, the location of the find and the levels of vegetation at the time in the area, Luticosaurus in this instance probably died in the Brook Bay area of the Isle of Wight through starvation. There was simply not enough vegetation for such a large animal to live on – especially as vegetation of the time was largely nutritionally poor and all herbivores had to consume large quantities on a daily basis of the tough, fibrous plant life.

Titanosaurus family members were slow-moving pack animals that would often move in large groups so the large adults could protect the more vulnerable young. If this was the case with the Brook Bay Luticosaurus, it raises the question of where the others that should have been travelling with the animal ended up. So far though, no other specimens of Luticosaurus have been found.

VITAL STATISTICS

NAME: Luticosaurus (loo-tee-ko-sore-us)

NAME MEANING: Jute Lizard

FAMILY: Titanosaur

ESTIMATED SIZE: 20m long

ESTIMATED WEIGHT: 26 tonnes

DIET: Herbivore

ANATOMICAL CHARACTERISTICS: Proportionally small head, long neck and tail, oval body shape with strong, evenly-spaced legs

LIVED DURING: Barremian stage of the Early Cretaceous

WHERE FOUND: Brook Bay, Isle of Wight, England

FOSSIL REMAINS: Three incomplete tail vertebrae

NEOVENATOR
NEW HUNTER

New Hunter was an aggressive carnivore which most likely fed on smaller dinosaurs like Hypsilophodon and perhaps even larger ones too like Iguanodon. The layout of the land in which Neovenator lived was largely open shrub land with mid-sized rivers flowing through. This means that rather than chasing prey down with a physically demanding pursuit, it was far more likely to be an ambush hunter.

While the lay of the land meant that there wasn't a lot of cover for the predator, it was equally difficult for the prey to stay out of the sight of big dinosaurs like Neovenator. Even though this dinosaur was very well evolved to hunt and rip apart prey quickly, it is entirely possible that a large part of its diet was made up of carrion and sick or injured animals, similar to the way in which wild lions survive today. Neovenator was a quick animal with rows of sharp, serrated teeth but it also looked every bit the aggressor thanks to horned extensions of the skull sited directly above its eyes.

It was also the first Allosauroid dinosaur to be discovered in Europe and fossils from this fearsome animal have been vcomparatively easy to come by on the Isle of Wight.

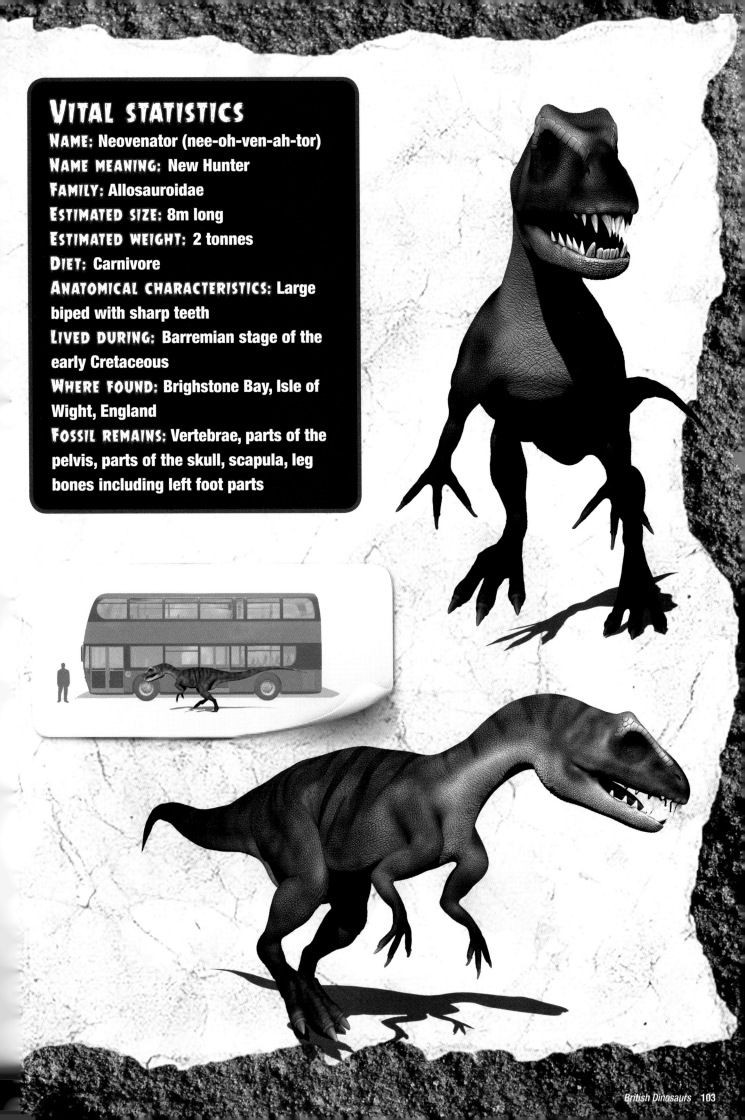

VITAL STATISTICS

NAME: Neovenator (nee-oh-ven-ah-tor)

NAME MEANING: New Hunter

FAMILY: Allosauroidae

ESTIMATED SIZE: 8m long

ESTIMATED WEIGHT: 2 tonnes

DIET: Carnivore

ANATOMICAL CHARACTERISTICS: Large biped with sharp teeth

LIVED DURING: Barremian stage of the early Cretaceous

WHERE FOUND: Brighstone Bay, Isle of Wight, England

FOSSIL REMAINS: Vertebrae, parts of the pelvis, parts of the skull, scapula, leg bones including left foot parts

OPLOSAURUS
ARMOURED LIZARD

Oplosaurus was a long-necked sauropod dinosaur that spent its time eating huge amounts of vegetation. Wandering slowly though the woodlands of the time, this large animal had to consume vast amounts of the nutritionally-poor food source just to survive.

Its teeth, therefore, were very similar to those of Brachiosaurus and it's likely that Oplosaurus shared many physical characteristics with its more well-known cousin.

All of these types of animals moved and lived in herds or super herds, travelling in gigantic groups from forest to forest in order to feed.

As with the earlier Cardiodon, so far the only fossil of Oplosaurus to be discovered is a single tooth. Since this is such a small part of the dinosaur and, at the time of writing, there are no other similar animals of this direct family discovered in the UK, it is entirely plausible that this was a lone animal that got lost and that its remains were washed into the Isle of Wight during a flood.

Its name probably comes from the Greek 'hoplon' which means armour or shield.

VITAL STATISTICS

NAME: Oplosaurus (op-low-sore-us)
NAME MEANING: Armoured Lizard
FAMILY: Sauropoda
ESTIMATED SIZE: 18m long
ESTIMATED WEIGHT: 15 tonnes
DIET: Herbivore
ANATOMICAL CHARACTERISTICS:
Comparatively small head, long
neck, long tail, equal squat legs,
round oblong body
LIVED DURING: Barremian stage of
the early Cretaceous
WHERE FOUND: Wealden Clay
near Brighstone Bay, Isle of Wight,
England
FOSSIL REMAINS: One tooth

ORNITHODESMUS
BIRD LINK

A small dinosaur, it's likely that Ornithodesmus would have been quick on its feet to hunt small lizards and animals that it could catch in the open shrub lands and low wooded areas of its habitat.

The ability to make quick, darting movements and pinpoint its prey with sharp eyesight were the most important attributes possessed by this small meat-eater.

When it was first discovered as a fossil, palaeontologists believed Bird Link was in fact a bird. Then it was reclassified as a pterosaur before it was finally realised that it was actually a feathered, terrestrial dinosaur.

Although it could not fly, Ornithodesmus was small, agile and fast; and while it may have been less than 2m long it was still a formidable animal that was typical of the clever and astute raptors of the time.

Feathers, long splayed fingers and a comparatively light build place this unusual predator at the crossroads between prehistoric dinosaur and modern bird.

VITAL STATISTICS

NAME: Ornithodesmus (or-nith-oh-dez-muss)

NAME MEANING: Bird Link

FAMILY: Dromaeosauridae

ESTIMATED SIZE: 1.8m long

ESTIMATED WEIGHT: 50kg

DIET: Carnivore

ANATOMICAL CHARACTERISTICS: Small and bipedal with primitive feathers and long snout resembling beak, small sharp teeth and bird-like eyes

LIVED DURING: Barremian stage of the Early Cretaceous

WHERE FOUND: Brook Bay, Isle of Wight, England

FOSSIL REMAINS: Six pieces of backbone which are all fused

ORNITHOPSIS
BIRD LIKENESS

'Bird Likeness' is now something of a misnomer for Ornithopsis as this was an animal classified originally as one thing and then identified as something else entirely. At first, when the vertebrae of this dinosaur were discovered by Harry Govier Seeley in 1870, Ornithopsis was thought to have been a halfway house between pterosaurs, birds and dinosaurs. However, this was rejected and later identification of the fossils showed that this was in fact a member of the Brachiosaur family.

So far, there have only been a few pieces of this dinosaur found and because this is a lone specimen it cannot be determined whether it was a solitary animal or a member of a larger herd.

Brachiosaurs were voracious herbivores and needed a lot of vegetation on a daily basis to maintain their huge bodies. A large group of them could wreak havoc on woodland areas and surrounding plant life.

Given what is known about the Isle of Wight during the Barremian stage of the Early Cretaceous, it is unlikely that this animal could have lived there for long. In fact it is entirely likely that there was little in the way of vegetation of the type that this Brachiosaur needed and it could have starved to death.

In an area of low-lying vegetation, competition between this tall animal and smaller herbivores would have placed it at a serious disadvantage.

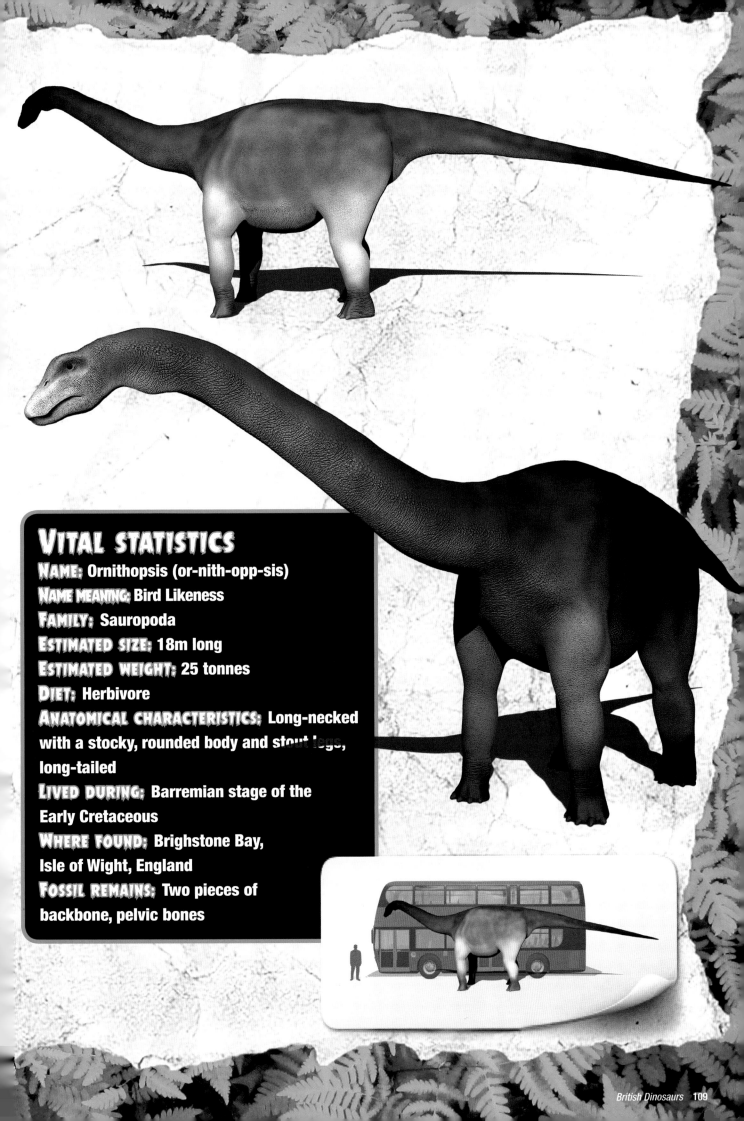

VITAL STATISTICS

NAME: Ornithopsis (or-nith-opp-sis)

NAME MEANING: Bird Likeness

FAMILY: Sauropoda

ESTIMATED SIZE: 18m long

ESTIMATED WEIGHT: 25 tonnes

DIET: Herbivore

ANATOMICAL CHARACTERISTICS: Long-necked with a stocky, rounded body and stout legs, long-tailed

LIVED DURING: Barremian stage of the Early Cretaceous

WHERE FOUND: Brighstone Bay, Isle of Wight, England

FOSSIL REMAINS: Two pieces of backbone, pelvic bones

POLACANTHUS
MANY SPINES

Polacanthus was an ankylosaur, a type of dinosaur that walked on four legs and had huge chunks of armour over its hips, neck, back and majority of its tail.

With back legs that are longer than those of other ankylosaurs, Polacanthus would have assumed a somewhat 'head-down' stance – but that didn't hinder it at all.

It had a large pelvic shield which was a fused, large bone that sat over its hips but effectively 'floated' because it wasn't attached to the main skeletal structure underneath. The skin was atypically thick with large armoured nodules over the shoulders, back, neck and tail and in places the plates of skin were completely fused to form tough areas of armour.

There were also several different lengths of spikes on Polacanthus, varying in size from its head back down the body. It would have been fairly slow-moving due to the weight of its armour, although it was a powerfully built animal. That weight would also have robbed it of any chance it might otherwise have had to dodge or evade the multitude of fast-moving agile predators of its time. Toughness and strong defences were its only hope for survival.

VITAL STATISTICS

NAME: Polacanthus (po-la-kan-thuss)

NAME MEANING: Many Spines

FAMILY: Polacanthidae

ESTIMATED SIZE: 4-5m long

ESTIMATED WEIGHT: 1-1.5 tonnes

DIET: Herbivore

ANATOMICAL CHARACTERISTICS: Quadruped with armour and spikes

LIVED DURING: Barremian stage of the Early Cretaceous

WHERE FOUND: Rudgwick, West Sussex, England

FOSSIL REMAINS: Backbone, leg bones, rib fragments and a partial shoulder

THECOCOELURUS
SHEATHED HOLLOW TAIL

Looking like a large flightless bird, Thecocoelurus had a small head with a short toothless beak similar to that of a parrot. Its arms had long talons and feathers which may have been for display and courtship.

Although most of its body was covered with primitive fur-like feathers, its scaly tail more closely resembled that of other meat-eating dinosaurs. It had long powerful legs with short strong toes well suited to moving at high speed.

While Thecocoelurus didn't have teeth it probably fed on smaller reptiles and insects that it picked away from the foliage of the British floodplains and lowlands during the Early Cretaceous period. It may have been omnivorous – using its arms to pull down branches to find green shoots.

Too little remains of Theocococelurus to be certain, but it may also have had a crest of some sort on top of its head as some other similar animals did.

The single piece of bone from which it is known was found by the Reverend William D Fox during the 1880s. It was eventually identified as an oviraptorosaur in 2001 but then reclassified as an ornithomimosaur in 2014.

VITAL STATISTICS

NAME: Thecocoelurus (theek-oh-seel-oh-russ)

NAME MEANING: Sheathed Hollow Tail

FAMILY: Ornithomimosauria

ESTIMATED SIZE: 1.5-2m long

ESTIMATED WEIGHT: 20-40kg

DIET: Carnivore

ANATOMICAL CHARACTERISTICS:
Feathered biped with large
wings and large neck,
small head, toothless and large
extended tail, feathered

LIVED DURING: Berriasian to Aptian eras of the
Early Cretaceous

WHERE FOUND: Brook Bay, Isle of Wight, England

FOSSIL REMAINS: Half a single piece of
backbone which are all fused

TITANOSAURUS
TITANIC LIZARD

It is perhaps unsurprising that Britain has Titanosaurus among its dinosaur ranks because these giants walked all over the earth.

Titanosaurus was a very large animal. From the tip of its long, whip-like tail to its comparatively short (for this type and family of dinosaur) neck this animal could easily get to be 12-15m long and would tip the scales at 13 tonnes.

Its forelimbs were longer than the rear ones and this gave Titanosaur a 'raised front' stance with a graceful gait as it walked.

This dinosaur lived and foraged in herds. It has been found that palms and conifers made up its diet which would have taken almost constant grazing with its small, evenly spaced teeth which looked something like an upturned garden rake.

It's almost certain that this was not an aggressive animal, despite the somewhat frowning appearance its nasal ridges on its long, narrow head would have given it. Outwardly, Titanosaur was covered in a mosaic of small scales which served to make it look like one huge mass of tough skin. It would have been very difficult for a predator of the time to get through to the flesh beneath during an attack.

Titanosaur mothers could be incredibly delicate, despite their huge bulk. When it came time to lay their 5in diameter eggs the mothers would dig a hole with their back legs and lay around 25 eggs at a time and then gently cover them with dirt and foliage.

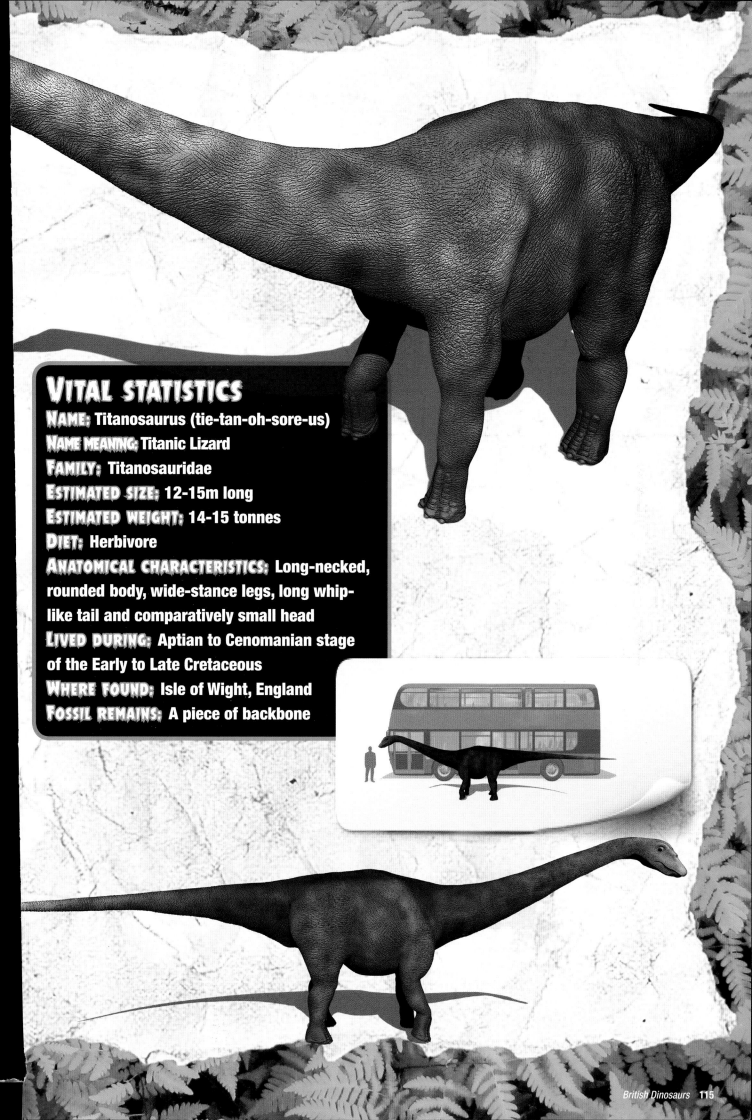

VITAL STATISTICS

NAME: Titanosaurus (tie-tan-oh-sore-us)

NAME MEANING: Titanic Lizard

FAMILY: Titanosauridae

ESTIMATED SIZE: 12-15m long

ESTIMATED WEIGHT: 14-15 tonnes

DIET: Herbivore

ANATOMICAL CHARACTERISTICS: Long-necked, rounded body, wide-stance legs, long whip-like tail and comparatively small head

LIVED DURING: Aptian to Cenomanian stage of the Early to Late Cretaceous

WHERE FOUND: Isle of Wight, England

FOSSIL REMAINS: A piece of backbone

VALDOSAURUS
WEALDEN LIZARD

It may have walked and even run on two legs with its head up and its long tail stretched out behind it like a meat-eater chasing down prey, but Valdosaurus was actually a herbivore.

It had smooth, flat teeth which would make short work of the tough vegetation but it had no means of defending itself other than speed – no plates of tough bone, no hardened nodules or thick skin. Putting off predators meant being fleet of foot to get out of the way of trouble, despite its 4m length.

Processing the fibrous and nutritionally poor plant matter of the time meant that Valdosaurus also needed to have a fairly enormous gut. Though it has been suggested that Valdosaurus was 'gazelle-like' in its movements, running about to avoid attackers with a bulky body must have been difficult and there is evidence to suggest that many did not manage to get away. Contemporary predators such as Eotyrannus almost certainly fed on Valdosaurus since their fossils have been found together.

It is very likely that Valdosaurus lived in herds, similar to Hypsilophodon. It was a primitive form of Iguanodon.

VITAL STATISTICS

NAME: Valdosaurus
(vall-doe-sore-us)

NAME MEANING: Wealden Lizard

FAMILY: Dryosauridae

ESTIMATED SIZE: 3-4m long

ESTIMATED WEIGHT: 40kg

DIET: Herbivore

ANATOMICAL CHARACTERISTICS:
Small and bipedal with
sharp teeth

LIVED DURING: Valanginian
stage of the Early Cretaceous

WHERE FOUND: Chilton Chine,
Isle of Wight and Hastings Beds
of West Sussex, England

FOSSIL REMAINS: Thigh bones,
broken skull fragments, partial
lower jaw and some teeth

VECTISAURUS
ISLE OF WIGHT LIZARD

Though it was able to walk on two legs, this dinosaur would have travelled on four legs as it roamed in herds seeking fresh greenery to eat.

The forearms of Vectisaurus were almost legs but ended in very distinctive hands. Two thumbs were positioned on opposite sides of the palm with the uppermost thumb getting a long spike of bone jutting out from the end. Vectisaurus also had thick skin with thicker areas around the back of the neck, shoulders and hips to help fend off attacks from predators.

Looking very similar to Iguanodon though smaller and more delicately proportioned, it has been suggested that the fossil remains known as Vectisaurus might actually be those of a young Mantellisaurus.

It has also been suggested that Vectisaurus, if not a Mantellisaurus, might have had a noticeable ridge along its back, giving it a more distinct appearance. Whatever the case, it is certain that Britain at this time was well populated by herbivorous animals that belonged to the same family as Iguanodon or looked very much like it.

It was named by John Whitaker Hulke in 1879.

VITAL STATISTICS

NAME: Vectisaurus
(vek-tee-sore-us)

NAME MEANING: Isle of Wight
Lizard

FAMILY: Iguanodontidae

ESTIMATED SIZE: 4m long

ESTIMATED WEIGHT: 1 tonne

DIET: Herbivore

ANATOMICAL CHARACTERISTICS:
Large and bipedal with flat,
chisel-like teeth

LIVED DURING: Barremian stage
of the Early Cretaceous

WHERE FOUND: Brighstone Bay,
Isle of White, England

FOSSIL REMAINS: Pieces of
backbone, teeth, small bits

YAVERLANDIA
FROM YAVERLAND POINT

One of the more frightening looking British dinosaurs was Yaverlandia. It was relatively small but had long pointed jaws lined with sharp little teeth.

Standing on two legs, it was covered with primitive feathers which in practice would have looked more like fur. Yaverlandia was quick and agile, able to use its long clawed fingers to grip its prey or to root through carrion remains seeking morsels to eat.

It had a relatively short neck and tail, however, suggesting that it was not built for running and speed. The precise purpose of its feather coat is unknown – but it may have been for insulation or for display. While its arms may appear to be almost vestigial wings, Yaverlandia was certainly not able to take flight.

The single piece of fossilised skull bone from which Yaverlandia is known was discovered during the late 1920s and was thought to belong to Vectisaurus. More than 50 years later, in 1971, palaeontologist Peter Galton realised that it belonged to a distinct species in its own right and named it Yaverlandia.

Galton thought it was a thick-headed Pachycephalosaur, however, and it was only recently that it was reclassified as a maniraptoran – a feathered meat-eater.

VITAL STATISTICS

NAME: Yaverlandia (yar-ver-lan-dee-ah)

NAME MEANING: From Yaverland Point

FAMILY: Maniraptora

ESTIMATED SIZE: 1m long

ESTIMATED WEIGHT: 10kg

DIET: Carnivore

ANATOMICAL CHARACTERISTICS: Small and bipedal with sharp teeth and feathers

LIVED DURING: Barremian stage of the Early Cretaceous

WHERE FOUND: Yaverland Point, Isle of Wight, England

FOSSIL REMAINS: Partial skull bone

MANTELLISAURUS
MANTELL'S LIZARD

A member of the Iguanodontidae family, Mantellisaurus was a slightly smaller version of the famous Iguanadon dinosaur family line. As a herbivore able to reach high up and low down, it roamed southern Britain eating a variety of plants including conifers, cycads and tree-ferns.

Mantellisaurus' head was hourglass shaped if viewed from above because of its expanded beak and this indicates that, along with the 40mm high teeth, it spent a long time grinding fibrous plant matter down in a grazing manner.

As with all members of the Iguanodontidae family, Mantellisaurus had dextrous hands with a short thumbspike and elongated metacarpals with three phalanges in the fourth and fifth fingers. Because the forearms were only 50% of the length of the legs, it appears that while Mantellisaurus was bipedal it often walked on all fours as a quadrupedal animal, with its head lifted up and its tail used as a counterweight. Mantellisaurus was a social creature living and moving in family groups.

There were large tendons visible under the Mantellisaurus' skin which gave the dinosaur something of a hunchbacked-look from the side.

It was named after famous early Victorian palaeontologist Gideon Mantell.

VITAL STATISTICS

NAME: Mantellisaurus (man-telly-sore-us)

NAME MEANING: Mantell's Lizard

FAMILY: Iguanodontidae

ESTIMATED SIZE: 7m long

ESTIMATED WEIGHT: 750kg

DIET: Herbivore

ANATOMICAL CHARACTERISTICS: Large bipedal with short arms, upright neck and long tail

LIVED DURING: Aptian stage of the Early Cretaceous

WHERE FOUND: Atherfield village, Brighstone Bay, Isle of Wight, England

FOSSIL REMAINS: Many examples including several almost complete skeletons

ENALIORNIS
BIRD OF THE SEA

With its short wing-like arms, light build, long beak and feathers, Enaliornis is one of the earliest dinosaurs to also be regarded as a bird – like Archaeopteryx.

Although would have been unable to fly, Enaliornis was able to dive through water to catch fish like a modern grebe or loon. Its body was slender and streamlined, its neck was long and flexible and its mouth was well suited to the task of snatching its fast-moving prey.

Unlike modern semi-aquatic birds, Enaliornis may not have had webbed feet however and its beak was lined with small sharply pointed teeth.

In addition, its feathers were less well developed than those of modern birds and would have given it a slightly fuzzy appearance.

While its long legs would have enabled Enaliornis to swim, it would also have been able to wade swiftly through the shallows of the rivers and ponds of its habitat in search of easy prey.

Many pieces of Enaliornis have been found in the area around Cambridge, but most have been badly fragmented and weathered. It was originally named Pelagornis – 'Open Sea Bird' – in 1866 but it was later found that this name was already taken by a sort of pelican from the much later Miocene period.

VITAL STATISTICS

NAME: Enaliornis (en-al-ee-or-nis)

NAME MEANING: Bird of the Sea

FAMILY: Enaliornithidae

ESTIMATED SIZE: 1m long

ESTIMATED WEIGHT: 3kg

DIET: Carnivore

ANATOMICAL CHARACTERISTICS: Short wing-like arms, long neck and long beak with teeth

LIVED DURING: Late Albian stage of the Early Cretaceous

WHERE FOUND: Upper Greensand, near Cambridge, England

FOSSIL REMAINS: Fragments including pieces of skull, arms, legs and backbone

ANOPLOSAURUS AND ACANTHOPHOLIS
UNARMOURED LIZARD AND SPINY SCALES

Although they differed in appearance, Anoplosaurus (below left) and Acanthopholis (left) both lived in what is now the Cambridgeshire area at around the same time. Both were low-browsing plant-eaters with armour plates on their backs, spikes for protection and beak-like mouths for cutting through the rough foliage they liked to eat.

As members of the Nodosauridae family, neither of them had a bony club at the end of its tail, despite otherwise being of the general ankylosaur shape. Anoplosaurus was the taller of the two, with a narrower body, longer legs and a more upright stance. Acanthopholis was wider and lower, and each had a different arrangement of bony armour plates and spines on its body.

During the latter part of the Early Cretaceous, when these animals lived, Cambridge was on the coast of a large landmass and some of the surviving fragments of Acanthopholis show evidence of having been gnawed on by sharks – presumably after its body had been washed out to sea by river waters. Exactly what predators might have been around to eat these grazers is unknown since too few fossils from the same period have survived.

Anoplosaurus (below right) was named in 1879, Acanthopholis (below left) in 1867. Today, there is uncertainty about which fossils actually belong to Acanthopholis but it is certain that Britain was inhabited by several species of ankylosaur at this time.

VITAL STATISTICS

NAMES: Anoplosaurus (Ah-nop-lo-sore-us) and Acanthopholis (ah-can-thoff-oh-liss)

NAME MEANINGS: Unarmoured or Unarmed Lizard and Spiny Scales

FAMILY: Nodosauridae

ESTIMATED SIZE: 3-5.5m long

ESTIMATED WEIGHT: 0.8-3 tonnes

DIET: Herbivore

ANATOMICAL CHARACTERISTICS: Slim with armour and spikes but no tail club

LIVED DURING: Late Albian stage of the Early Cretaceous

WHERE FOUND: Various parts of Cambridgeshire and Kent, England

FOSSIL REMAINS: Parts of skull, armour plates and shoulder blade for Anoplosaurus; parts of legs, backbone and other pieces for Acanthopholis

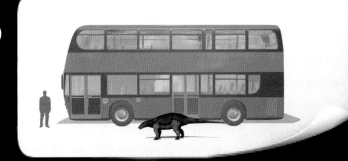

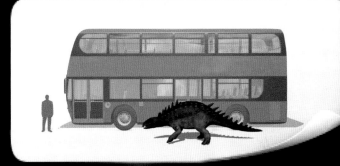

EUCERCOSAURUS
GOOD TAILED LIZARD

Roaming the coastal areas of what is now central Cambridgeshire at the same time as Anoplosaurus and Acanthopholis was the similar-looking Eucercosaurus.

Another low-browser of tough vegetation, this armoured animal belonged to a different branch of the ankylosaur family and had a bony club at the end of its tail for warding off predators.

As mentioned earlier, exactly what those predators were is unknown but Neovenator may have shared the same time period as well as another species, an allosauroid, found not too far away in France called Erectopus 'Upright Foot'.

Faced with Eucercosaurus's powerful tail neither would have found it a very easy meal. Even if they managed to avoid its defensive blows, they would have encountered the difficult task of finding somewhere on its heavily armoured and spike covered body to sink their teeth into.

Its low-to-the-ground stance would have made is doubly difficult for attackers to find a weak spot.

Eucercosaurus was named in 1879 but is only known from a few pieces of spine. In recent times, it has been suggested that those fragments of fossilised bone might actually have belonged to an ornithopod instead – an animal that belonged to the same family as Iguanodon.

VITAL STATISTICS

NAME: Eucercosaurus (you-sir-co-sore-us)

NAME MEANING: Good Tailed Lizard

FAMILY: Ankylosauridae

ESTIMATED SIZE: 3-5m long

ESTIMATED WEIGHT: 1-3 tonnes

DIET: Herbivore

ANATOMICAL CHARACTERISTICS:
Standard ankylosaur form with tail club

LIVED DURING: Late Albian stage of the
Early Cretaceous

WHERE FOUND: Trumpington,
Cambridgeshire, England

FOSSIL REMAINS:
Pieces of backbone

TRACHODON
ROUGH TOOTH

Trachodon was a grazing plant-eater that could rear up on its long hind legs to reach foliage 4m above the ground. Its teeth had evolved so that it could mash them together with a movement similar to chewing – although it has been suggested that, unlike any modern animal, while its lower jaw remained fixed its flexible upper jaw did the moving.

It was a duckbilled dinosaur that came from the very end of the Early Cretaceous period and apart from its bill, being a large heavy animal, it shared very few characteristics with the modern water fowl. It had useful 'hands' for manipulating its food and a thick neck and body well suited to processing large amounts of plant matter.

One of the earliest members of the hadrosaur family, Trachodon was an evolutionary step on from the Iguanodon-type animals that had been so successful in Britain throughout the early Cretaceous.

Along with another hadrosaur-type, Iguanodon hillii, Trachodon is the most 'modern' known British dinosaur – even though all that remains of its entire species is a single tooth.

The loss of nearly all Britain's Late Cretaceous rocks through erosion means that unfortunately nothing is known about the animals that lived alongside Trachodon and Iguanodon hillii.

VITAL STATISTICS

NAME: Trachodon (trak-oh-don)
NAME MEANING: Rough Tooth
FAMILY: Hadrosauridae
ESTIMATED SIZE: 12m long
ESTIMATED WEIGHT: 8 tonnes
DIET: Herbivore
ANATOMICAL CHARACTERISTICS: Large duckbilled head, short neck, small forearms and longer rear legs
LIVED DURING: Late Albian to Early Cenomanian stage of the Cretaceous
WHERE FOUND: Cambridge area, Cambridgeshire, England
FOSSIL REMAINS: A single tooth